# WORK INJURY
## Management and Prevention

**Susan J. Isernhagen, PT**
Isernhagen and Associates
Duluth, MN

AN ASPEN PUBLICATION®
Aspen Publishers, Inc.      1988      Rockville, Maryland

Library of Congress Cataloging-in-Publication Data

Work injury: management and prevention/edited by Susan J. Isernhagen.
p. cm.
"An Aspen publication."
Includes bibliographies and index.
ISBN: 0-87189-788-1
1. Wounds and injuries—Treatment. 2. Industrial safety. 3. Physical therapy. I. Isernhagen, Susan J.
[DNLM: 1. Accidents, Occupational. 2. Physical Therapy. 3. Wounds and Injuries. Wa 485 P578]
RD97.5.P48   1988   615.8'2—dc19
DNLM/DLC
for Library of Congress
88-19334
CIP

Copyright © 1988 by Aspen Publishers, Inc.
All rights reserved.

Aspen Publisher, Inc., grants permission for photocopying for limited personal or internal use. This consent does not extend to other kinds of copying, such as copying for general distribution, for advertising or promotional purposes, for creating new collective works, or for resale. For information, address Aspen Publishers, Inc., Permissions Department, 1600 Research Boulevard, Rockville, Maryland 20850.

Editorial Services: Ruth Bloom

Library of Congress Catalog Card Number: 88-19334
ISBN: 0-87189-788-1

*Printed in the United States of America*

2 3 4 5

# Table of Contents

| | | |
|---|---|---|
| Contributors | | xiii |
| Preface | | xv |
| Acknowledgments | | xvii |
| Section 1— | **The Spectrum of Work Injury and Its Management** *Susan J. Isernhagen* | 1 |
| | Scope of the Field | 1 |
| | Work Injury: The Problem with Many Faces | 3 |
| | Conclusion | 6 |
| PART I— | **INJURY PREVENTION PROGRAMS** | 7 |
| Section 2— | **Health Education in the Workplace** *Margaret I. Bullock* | 9 |
| | Introduction | 9 |
| | Planning an Educational Program | 10 |
| | Conclusion | 18 |
| Section 3— | **Back Schools** *Susan J. Isernhagen* | 19 |
| | Introduction | 19 |
| | Are Back Schools Effective in Preventing Injury? | 19 |
| | What Is the Role of Fitness Instruction and Education in Prevention Programs? Is It Helpful? | 22 |

|  |  | What Would the Ideal Injury Prevention Program Contain? | 23 |
|---|---|---|---|
|  |  | What Are Some Challenges for the Future in Injury Prevention Programs? | 26 |
| Section 4— | | **Industrial Health Programs** | 29 |
|  | *Susan J. Isernhagen* | | |
|  | 1. | Fitness and Health Promotion Programs in Industry | 30 |
|  |  | Introduction | 30 |
|  |  | Fitness Works | 30 |
|  |  | Physical Fitness | 31 |
|  |  | Smoking Cessation | 32 |
|  |  | Hypertension | 32 |
|  |  | Stress Management | 33 |
|  |  | Weight Control | 33 |
|  |  | Summary | 33 |
|  | 2. | Pause Gymnastics—Exercise Breaks | 35 |
|  |  | Practical Aspects of the Program | 35 |
|  |  | Benefits of the Program | 37 |
| Section 5— | | **The NIOSH Strategy for Reducing Musculoskeletal Injuries** | 39 |
|  | *Timothy J. Pizatella, Roger M. Nelson, David E. Nestor, and Roger C. Jensen* | | |
|  |  | Scope of the Field | 39 |
|  |  | The NIOSH Strategy | 40 |
|  |  | Conclusion | 50 |
| PART II— | | **ERGONOMICS** | 54 |
| Section 6— | | **Ergonomics and Cumulative Trauma** | 54 |
|  | *Susan J. Esernhagen* | | |
|  |  | Introduction | 54 |
|  |  | Cumulative Trauma and Work Injury | 56 |

| | | |
|---|---|---|
| Section 7— | Matching Worker and Worksite—Ergonomic Principles ........................................................................ | 65 |
| | *Suzanne H. Rodgers* | |
| | Characterizing the Job Demands .............................. | 65 |
| | Characterizing the Person's Capabilities for the Work ............................................................................ | 72 |
| | Examples of Techniques To Reduce the Demands of Heavy Jobs ............................................................ | 74 |
| | Summary ........................................................................ | 77 |
| Section 8— | Three Approaches to Specific Ergonomic Problems .......................................................................... | 80 |
| | *Susan J. Isernhagen* | |
| | Repetition Strain Injury—Understanding Prevention and Management ................................ | 81 |
| | Ergonomic Intervention As Part of an Injury Prevention Program ............................................... | 85 |
| | Worksite Analysis ...................................................... | 86 |
| PART III— | PRE-EMPLOYMENT AND PREPLACEMENT SCREENING ...................... | 91 |
| Section 9— | Pre-Employment Screening: The Physical Perspective ................................................................. | 92 |
| | *Alan W. Morris and Charles K. Anderson* | |
| | Evaluation of Alternative Selection Procedures.......... | 93 |
| | Alternative Forms of Screening ................................ | 95 |
| | Development and Implementation of a Screening Program .................................................................... | 101 |
| | Conclusion .................................................................... | 103 |
| Section 10— | Preplacement Physical Examinations .................. | 106 |
| | *William P. Fleeson* | |
| | Problems with Traditional Medical Examinations ..... | 107 |
| | The Directed Occupational Medicine Examination ... | 107 |
| | Other Reasons for Doing Preplacement Physical Examinations ............................................................ | 109 |
| | Successful Screening Experience ................................ | 111 |

## PART IV— CARE AFTER INJURY ... 113

### Section 11— Immediate Care Delivery Systems ... 115
Susan J. Isernhagen

Recognition of Early Care ... 115
Components of a Delivery System for Care After Injury ... 116
Examples of Delivery Systems ... 117
"Perks" for Work Injured Patients ... 120
One System Explored ... 120
Prerogative of Industry in Medical Care Selection ... 124
Summary ... 125

### Section 12— Role of the Physical Therapist in Early Intervention ... 127
Susan J. Isernhagen

Introduction ... 127
Delivery Systems for Immediate Care of Industrial Injuries ... 128
Trends in Immediate Care ... 130
Types of Physical Therapy Interventions in Immediate Care ... 130
Return to Work ... 131
Trends and Future Challenges ... 132

### Section 13— Stabilization Principle for Physical Management of Work Injuries ... 134
Arthur H. White

## PART V— FUNCTIONAL CAPACITY EVALUATION ... 137

### Section 14— Functional Capacity Evaluation ... 139
Susan J. Isernhagen

Overview ... 139
Functional Capacity Evaluation Parameters ... 140
Design of the Functional Capacity Evaluation ... 144
Components of Functional Capacity Evaluation ... 148
Evaluator Perspective ... 149
Report Formats ... 162

Case Study—Functional Capacity Evaluation ............ 164
Adjuncts to Functional Capacity Evaluation ............. 164
Legal Implications ................................................... 166
Appendix 14-A—Functional Capacities Evaluation
  Case Study ........................................................... 168

1. Disability Determinations .................................... 175
   *Jack Casper*

   Social Security Disability Hearings ..................... 176
   Standardized Job Traits ...................................... 177
   Alternate Sit/Stand Jobs .................................... 179
   Discussion .......................................................... 179

2. Cost Savings in Four Cases ................................. 181
   *Margot Miller*

   Case Study #1 .................................................... 181
   Case Study #2 .................................................... 182
   Case Study #3 .................................................... 182
   Case Study #4 .................................................... 183

3. Functional Capacities Assessment Research: The
   Relationship of Age and Gender to Functional
   Performance—Patients and Uninjured Subjects ... 184
   *Susan J. Isernhagen, Karen Mokros,*
   *Margot Miller, and Laurie Johnson*

   Overview ............................................................ 184
   Introduction ....................................................... 185
   Purpose .............................................................. 185
   Content .............................................................. 186
   Professional Implications and Clinical
     Relevance ........................................................ 190
   Recommendations for Future Work .................... 191

**PART VI— WORK HARDENING** ............................ 193

**Section 15— Work Hardening** ................................. 195
*Carol Frantz Lett, Naomi Elizabeth McCabe,*
*Anne K. Tramposh, and Suzanne Tate-Henderson*

Introduction ............................................................. 195
Need for Work Hardening ........................................ 195

        Work Hardening: An Evolutional Process ................ 196
        Literature Review ....................................................... 196
        Types of Work Hardening Services ........................... 198
        An Example Program ................................................. 206
        Professional Application ........................................... 223
        Organizational Problems ........................................... 224
        Environmental Problems ........................................... 224
        Case Closure .............................................................. 225
        Summary .................................................................... 228

**PART VII— FUNCTIONAL RESTORATION OF THE PATIENT WITH CHRONIC SPINAL DISORDERS** ....................................................... 231

**Section 16—** Functional Restoration of the Patient with Chronic Spinal Disorders ................................ 232
*Tom G. Mayer*

        Is Measurement the Key to Rehabilitation of Spinal Disorders? ............................................................. 232
        How Can Efficient Spinal Measurements Be Designed? ................................................................ 234
        What Barriers to Functional Recovery May Arise? ...... 236
        Functional Restoration: Rebuilding the Deconditioned Mind and Body ............................ 237
        How Can These Concepts Be Summarized? ............... 242

**PART VIII—PSYCHOLOGICAL VARIABLES AND ISSUES** ................................................................ 245

**Section 17—** Psychological Barriers to Recovery ...................... 247
*Lynne E. Killian*

        Psychological Aspects of Delayed Recovery ............... 247
        Role of the Psychologist in Industrial Medicine Programs ............................................................... 251

**Section 18—** Symptom Magnification Syndrome ...................... 257
*Leonard N. Matheson*

        Background ................................................................ 257
        Developmental Origins .............................................. 258

Identification .............................................. 262
Evaluation .................................................. 269
Treatment .................................................. 274
Summary .................................................... 280

**Section 19— Chronic Pain Management** ................................ 283
*Jodi Landis West*

Work Injury and Chronic Pain .................................. 283
Chronic Pain Attributes ........................................ 284
History of Treatment Models .................................. 284
Program Planning .............................................. 286
Components of a Pain Management Program ........... 287
Pain Center Experience ....................................... 290

**PART IX— THE RETURN-TO-WORK PROCESS** ................ 295

**Section 20— Return to Work: The Physician's Role** ................. 298
*William P. Fleeson*

The Occupational Physician and the Company:
  Partners .................................................. 298
Correct Diagnosis: The Foundation of Correct
  Care ..................................................... 299
Beginning of the Treatment Plan .............................. 301
Transmitting Information to the Employer ................... 303
Ongoing Care .................................................. 304
Ending Care ................................................... 307
Special Problems Associated with Worker's
  Compensation ............................................. 308
Early Return to Work and Light Duty ......................... 309

**Section 21— The Rehabilitation Consultant** .......................... 311
*Suzanne Tate-Henderson and Timothy R. Johnson*

Introduction .................................................. 311
The Planning Function ......................................... 311
The Organizing Function ....................................... 313
The Leading Function .......................................... 314
The Controlling Function ...................................... 315
Summary ....................................................... 316

## Section 22 — Employer Point of View .................... 317
*Norma E. Swanbum*

Preventing the Injury .................................... 317
When the Injury Occurs ................................ 318
Restricted Duty Program ............................... 322
Communicating with Medical Providers ........... 323
Modifying a Job ........................................... 325
Peer Pressure from Co-Workers ...................... 325
Possible Peer Pressure from the Supervisor ...... 326
Labor Unions .............................................. 327
Communication ........................................... 328
Insurers ...................................................... 328
Summary .................................................... 329

## Section 23 — The Therapist ............................... 331
*Susan J. Isernhagen*

An Extension beyond the Medical Model ......... 332
Active Communication ................................. 332
Objective and Unbiased Position ..................... 332
Transferable Skills ....................................... 333
Follow-Up/Clinical Research ......................... 333
Therapist Roles in Return-to-Work Process ..... 334
Summary .................................................... 337

## Section 24 — The Attorney's Role ...................... 338
*Susan J. Isernhagen*

Introduction ................................................ 338
Views on the Workers' Compensation System and
   Legal Processes ........................................ 339
Perspectives on the Attorney's Role Regarding a
   Client ..................................................... 341
What is the Attorney's Role in Return to Work? .. 342
How Do You View the Role of Other Professionals
   in the Workers' Compensation System? ....... 344

## Section 25 — The Main Feature: The Employee ..... 347
*Margot Miller*

Able Clients ................................................ 347

|  |  | |
|---|---|---|
|  | Unable Clients | 349 |
|  | Fearful Clients | 350 |
|  | Unmotivated Clients | 352 |
|  | Summary | 353 |
| PART X— | **FUTURE CHALLENGES IN OCCUPATIONAL MEDICINE** | **355** |
|  | *Susan J. Isernhagen* |  |
|  | An Emerging Force | 355 |
|  | Commitment of Professionals | 355 |
|  | External Forces and Regulations | 356 |
|  | Changing Industrial Factors | 358 |
|  | Summary | 359 |
| **Index** |  | **361** |

# Contributors

**Charles K. Anderson, PhD**
Vice-President, Ergonomics and Engineering
Back Systems, Inc.
Dallas, Texas

**Margaret I. Bullock, PhD**
President
Academic Board
University of Queensland
St. Lucia, Queensland, Australia

**Jack Casper, MS, QRC**
Vocational Specialist
Crawford and Company
Duluth, Minnesota

**William P. Fleeson, MPH, MD**
Director, Industrial Medicine Center
Clinical Faculty, UMD Medical School
Associate Medical Director
Polinsky Medical Rehabilitation Center
Duluth, Minnesota

**Roger C. Jensen, PE**
Senior Engineering Officer
National Institute for Occupational Safety
 and Health
Division of Safety Research
Morgantown, West Virginia

**Laurie Johnson, PT**
Coordinator, FCA Network
Polinsky Medical Rehabilitation Center
Duluth, Minnesota

**Timothy R. Johnson, MS**
President
Vx: Vocational Expertise
Kansas City, Kansas

**Susan J. Isernhagen, RPT**
President
Isernhagen and Associates
Duluth, Minnesota

**Lynne E. Killian, MA**
Licensed Psychologist
Polinsky Medical Rehabilitation Center
Duluth, Minnesota

**Carol Frantz Lett, OTR, CIRS**
Vice President, Clinical Services
WX: Work Capacities, Inc.
Independence, Missouri

**Naomi Elizabeth McCabe, MS, OTR/L**
Work Capacities Services Coordinator
CHSA Network
Joliet, Illinois

**Leonard N. Matheson, PhD**
Licensed Psychologist
Director, Employment and Rehabilitation Institute of California
Anaheim, California

**Tom G. Mayer, MD**
Clinical Professor of Orthopedic Surgery
University of Texas Southwestern Medical Center
Medical Director
Productive Rehabilitation Institute of Dallas for Ergonomics (PRIDE)
Dallas, Texas

**Margot Miller, RPT**
Director, FCA Network
Polinsky Medical Rehabilitation Center
Duluth, Minnesota

**Karen Mokros, RN, PhD**
President
KLM Enterprises
Duluth, Minnesota

**Alan W. Morris, MS, LPT**
Executive Director
Network for Physical Therapy
Carrollton, Texas

**Roger M. Nelson, PhD**
Research Physical Therapist
National Institute for Occupational Safety and Health
Division of Safety Research
Morgantown, West Virginia

**David E. Nestor, PT, MS**
Research Physical Therapist
National Institute for Occupational Safety and Health
Division of Safety Research
Morgantown, West Virginia

**Timothy J. Pizatella, MS**
Chief, Accident Analysis Section
National Institute for Occupational Safety and Health
Division of Safety Research
Morgantown, West Virginia

**Suzanne H. Rodgers, PhD**
Consultant in Ergonomics/Human Factors
Rochester, New York

**Norma E. Swanbum**
Workers' Compensation Administrator
Minnesota Power
Duluth, Minnesota

**Suzanne Tate-Henderson, MS, CRC**
President
Vocational Economics, Inc.
Kansas City, Missouri and San Francisco, California
President: WX: Work Capacities, Inc.
Independence, Missouri

**Anne K. Tramposh, RPT, CIRS**
Director of Industrial Services
WX: Work Capacities, Inc.
Westwood, Kansas

**Jodi Landis West, PT**
Indiana Center for Rehabilitation Medicine
Indianapolis, Indiana

**Arthur H. White, MD**
Medical Director
San Francisco Spine Institute
Medical Director
Spine Care Medical Group
Seton Medical Center
Daly City, California

# Preface

People ask (and I ask myself), "How did you get involved in producing a book?" My answer lies in the need I, myself, had for a textbook of this type. Because occupational medicine is a relatively new field, those of us who develop work injury programs have a need to share ideas and learn from one another. By committing programmatic philosophies and principles to a written form, we can advance the dissemination of information and provide dialogue for the future.

The first purpose of this book was to describe successful management and prevention programs. Chapters were written by a wide variety of professionals because our work injury management effectiveness depends upon the team approach. Respect for, and communication of, points of view, needs, and competencies are critical to our success. The integration of medical, industrial, and legal points of view into one body of knowledge will reinforce the integration of philosophies in actual case management.

The second purpose for the text was to provide an educational springboard for direction and implementation of quality systems. For example, in the medical field, most schools do not currently teach functional capacity evaluation, work hardening, ergonomics, or other industrial medicine programs. Physical therapists, occupational therapists, psychologists, physicians, vocational counselors, and employers all need an educational model for implementation of these systems. Progress, choice, or change are best made when there is a theoretical or practical model from which to work.

The book begins with structured injury prevention techniques, progresses into early intervention of injury, the evaluation and treatment of chronic problems, and then presents challenges for the future. In addition, specific programs and professions are highlighted. The contributors to the text are knowledgeable, gracious, and (not surprisingly) unified in their philosophies. They are highly regarded not only in their own fields, but across disciplines.

The threads running through the chapters are easily perceived:

- attention to quality in professional programs
- emphasis on the worker as an important person, and on the worker's problems as priorities
- appreciation of the team approach
- importance of early return to work after injury
- necessity of injury prevention
- dedication to making work injury management even more effective in the future.

I believe all of the authors and contributors join me in asking you, the reader, to provide feedback on the philosophies and programs that we present here. The management of work injury has made remarkably positive progression in the past years, and we all will need to work together to continue that progression into the future.

*Susan J. Isernhagen*
*August 1988*

# Acknowledgments

I gratefully acknowledge the assistance and support I received during the production of this book.

I thank Dennis Isernhagen for strong professional and personal support throughout the entire process. His ideas and encouragement facilitated the work on this book. I also thank my daughters, Aura and Jessica, for their assistance in production aspects and their caring attitude.

I also gratefully acknowledge co-workers who have encouraged the pursuit of professional goals: Margot Miller for her support, philosophical sharing, and mutual dedication to quality programs; Mary Beaupre for her tireless and efficient typing, proofing, and suggestions for excellence in the written word.

Lastly, I would like to thank all authors for their excellent contributions. I have learned from them and it has been my pleasure to interact with these leaders in our professions.

SJI

# Section 1

# The Spectrum of Work Injury and Its Management

*Susan J. Isernhagen*

## SCOPE OF THE FIELD

Since the early 1980s, there has been an increased awareness of the need for effective programs in occupational medicine. In response to this need, referrals for programs dedicated to effective work injury management have been generated, sometimes ahead of their adequate development by medical facilities. Education in the government and business, manufacturing, medical, and insurance industries has been directed toward programs that prevent or reduce work injury or that rehabilitate employees for a safe, timely return to work. This awareness by all parties has led to the creation and growth of an industry—occupational health and medicine.

For the people who are dedicated to occupational health and medicine, there is a strong need for more program development. Professional areas, being comprehensive and diverse, are beginning to interlock programs rather than to compete.

Dean Imbrogno, MD, of Med Work, Dayton, Ohio,[1] has developed a working definition structure for occupational medical care. The following definitions are condensed and adapted from that structure.

### Occupational Health

Imbrogno[1] categorizes occupational health as embracing the multitude of health-related programs in all aspects of industrial effects on the body. Included in (but not limited to) this broad definition would be the specialties of:

- disaster plans and protocols
- poisonous substance control and treatment
- ergonomics
- industrial environmental hazards, such as dust, air pollution, and radiation
- governmental regulation compliance
- occupational medicine

**Occupational Medicine**

As a subgroup of occupational health, Imbrogno[1] further defines occupational medicine as specifically covering employee physical health and management of injury. Categories in occupational medicine would include (but not be limited to):

- physical examinations
- wellness and fitness programs
- work injury management
- rehabilitation
- injury prevention programs

**Physical Work Injury Management**

Going beyond those two definitions, this text falls categorically into the working definition of occupational medicine, but the field has been specialized further. The objective of this text is to delineate programs and philosophies that involve management of physical work injury. Prevention and rehabilitation of musculoskeletal conditions are the primary focuses of the chapters. The well-rounded approach to physical work injury includes (but is not limited to):

- physical injury prevention programs
- ergonomic intervention
- physical screening before work
- early, effective care of injury
- functional evaluation after injury
- worker rehabilitation—physical and psychological components

## WORK INJURY: THE PROBLEM WITH MANY FACES

Often the case of the injured worker is seen as only a one-faceted problem by the parties involved. As in most things in life, problems are solved according to the way they are perceived by the solver; ie,

- The medical specialists want to solve the problem by medical intervention, which makes the patient better.
- The employer wants to solve the problem by keeping the job intact and having the employee return to work.
- The insurance company wants to solve the problem by quick medical (cost) resolution and no disability liability.
- The employee just wants to return to normal (which usually includes full return to work).

For effective participation in work injury management one must view the whole picture.

Anyone working with workers' compensation clients/patients has most likely heard more than one version of a single work injury case. The versions were given from the perception of the participants' own interpretation. This points out the role of "point of view" in the working system. The problems that are inherent in the work injury management system are often the result of lack of understanding of other points of view.

Although this can happen in any medically oriented problem, it is particularly difficult with work injury because of the background of the professionals involved. There is a melding of medicine, industry, insurance, and government. The language and goals are significantly different. Past adversarial positions or a history of unsatisfactory results also may color present or future positions.

### The Players

To successfully participate in the workers' compensation system, one must first recognize that it is, in fact, a system, and second that each participant plays a different role in the system.

The employee and employer are the basic "players" in the workers' compensation game. To avoid adversarial tones in this relationship, the workers' compensation system was designed. It was intended to be "no fault" and thus simplify payment of benefits to injured workers. If, indeed, there is no fault and return to work is everyone's acknowledged goal, this system might work smoothly. However, as in all systems meant to simplify, human and situational variables create complexity. Thus, many relationships in the system are not always conducive to return to work, and at times each party might feel underinformed and not in control.

*Employee*

The employee, who should be the prime focus of the system, is often the one that knows the least and has less active participation than the attending professionals. Paychecks stop and are replaced by a reimbursement system, which is less predictable in both value and timeliness. The physician takes care of the injury; but, because reimbursement is directly related to disability, there may be a conflict between wellness and guaranteed compensation. In more complicated cases, medical diagnosticians or treatment personnel may give conflicting views on the employee's wellness related to the ability to return to work. In some instances, there are medical accusations regarding malingering that attack not only the employee's physical condition but also his or her personal integrity.

*Employer*

Second to the employee, the employer is the next affected party in a work injury. A productive person in the work system is now missing; and the employer, as owner of the work and the workers' compensation insurance, is responsible. Not only is there usually regret for the injury, but the associated difficult aspects are compounded by the need to temporarily replace the worker, to validate the cause of the accident, to file insurance claims, and, then, to wait for medical information regarding the return to work. The employer tends to be optimistic, seeking an early return to work to solve the difficulties of the work injury problem. Also, an early return to work alleviates anxiety regarding potential permanent injury of the employee.

Without the benefit of a medical background there is quite a burden on the employer. Often there is either no medical report forthcoming or the report is riddled with medical jargon that fails to address the work function of the employee. Employers are all too familiar with lengthy periods off work and subsequent suspicions that arise if the injury was subtle. Also, preparing coworkers and foremen for the injured employee's return to work often requires preplanning, diplomacy, and a positive attitude toward a successful transition. If work or worksite modifications are requested by the therapist, physician, or rehabilitation consultant, additional adaptability is needed. The employer needs support and communication but often is poorly informed about medical cause and prognosis.

### Medical Professionals

Historically, medical care providers operated in the spectrum of medical exclusivity. Physicians were initiators and coordinators of medical components. Other medical providers, including therapists, received referrals, proceeded with their work, and ended their involvement when treatment ended.

Medical professionals are hampered as work-injury managers because of both their education and their past/current practice. The physician, who is often the main link to other professional areas, generally is so involved in being a health care provider that acting as a coordinator is impossible. In many cases, the most valuable component (ie, the functional information) is known by the therapist (physical or occupational) but not necessarily transmitted to the physician or employer.

When the return-to-work process is studied, the communication link between medicine and industry is often viewed as the weakest.

### Rehabilitation Consultant

When the out-of-work period becomes prolonged or the return-to-work outcome is in jeopardy, other professionals are there to "help." The most recent advancement in occupational medicine is the development of the rehabilitation consultant (nurse, vocational specialist, etc). The rehabilitation consultant may work for an independent company or an insurance company, or may be affiliated with the state workers' compensation in the role of coordinator, counselor, and professional consultant. This person integrates medical, vocational, and specific worksite information and interfaces directly with the employer, employee, and medical specialists. When the system works, the rehabilitation consultant (case manager) facilitates communication and the return-to-work process. By design, this position should be neutral.

### Governmental Agencies

State workers' compensation system administrators have recently been under fire to reduce workers' compensation costs. High workers' compensation premiums cause problems not only for the employer but also for the state. The overall effect is a high cost of doing business in the state, leading to a poor business climate. As is usual in political situations, legislators and workers' compensation administrators are pressured to "change" the system. Pressure from employers to cut workers' compensation costs by reducing benefits or settlements is countered by pressure from union and employee groups who promote high benefits and settlements. Each group is rightfully protecting its own interests, but overprotectiveness has sometimes led to the development of confusing, contradictory, and inefficient laws that govern the workers' compensation system.

A recent direction of cost containment is to put a ceiling on reimbursement for medical evaluation and treatment. Although this sounds logical at first, it has often prevented the best medical care from being used and has perpetuated a greater use of low-cost, low-quality programming. In the long run, money spent on early quality care will be more effective than extensive dollars if the case becomes chronic.

*Attorney*

Out of the mistrust, confusion, or even abuse within the system the role of the attorney has arisen. Because the attorney's role is to support the client (whether employee or employer), the resultant cases may be determined by the strongest one-sided testimony. Financial case resolution is the attorney's goal, but often left behind are bitter feelings and even perceived dishonesty by one side or both sides. Can it really be helpful to future work function if an employee has "proved" that he or she is work disabled? Conversely, what happens to a work future for an employee who has been shown to be exaggerating an injury claim for money or to avoid work? Litigation, although very needed in some cases to protect the parties, often is seen as a detriment to the return-to-work process.

## CONCLUSION

This text is written by experts in rehabilitation, medicine, industry, workers' compensation, and law; and, although each has a recognized specialty, the reader will find among these authors a recognition of the need for a multiprofessional approach. The professional interrelationships are interwoven into the fabric of the text, promoting a unified approach.

**REFERENCE**

1. Imbrogno D: Practice diversification, in American College of Emergency Physicians (eds): *Marketing and Diversification Course*. New York: American College of Emergency Physicians, 1987.

# PART I

# Injury Prevention Programs

The sections in Part I explore specific programs designed to address (1) health and (2) injury prevention. Many of these interventions will also directly improve health and reduce injury in the workers' daily lives. Incorporation of these basic concepts may lead to an improvement in health for entire families.

### Section 2  Health Education in the Workplace

A comprehensive view of education in ergonomics and safety is presented. It forms the basis of multidimensional planning and implementation of injury-prevention programs. Ideas presented here will also be explored further in Part II, Ergonomics.

### Section 3  Back Schools

The most prevalent and well-accepted prevention programs in industry are those designed for back safety. Related in this section is professional evidence that education is preventive in nature. An educated employee is the first line of defense against work injury.

### Section 4  Industrial Health Programs

1. Fitness and Health Promotion Programs in Industry

With increased cultural emphasis on total fitness, employers as well as the general public are significantly more interested in health through fitness. Employers can enable their employees to actively engage in good health practices. The benefits translate to the worksite in the forms of decreased health insurance and workers' compensation costs.

2. Pause Gymnastics—Exercise Breaks

An extension of the physical fitness theme at work is exercise to reduce injury. Specific work causes specific musculoskeletal problems and proper exercise can be preventive.

## Section 5  The NIOSH Strategy for Reducing Musculoskeletal Injuries

Many world governments participate in setting guidelines and enforcing rules designed to reduce work injury. In a recent statement, the U.S. governmental agency, the National Institute for Occupational Safety and Health set forth a structural plan to incorporate data, research, and future work design into a systematic approach. The components of the plan are worthy of review here, not only as an indication of a governmental interest but also as a design for many injury-management programs.

# Section 2

# Health Education in the Workplace

*Margaret I. Bullock*

## INTRODUCTION

In an attempt to achieve the objectives of safe work practices, professionals have focused on job design and improvement of work methods. This has often involved a redesign of tasks, machine, or environment. They also have considered the feasibility of preselecting the worker, and the introduction of principles of job rotation, job enlargement, and job enrichment. In addition, a need for well-planned programs of education for all concerned with work has been realized, and increased emphasis has been placed recently on this aspect of prevention within the overall ergonomic approach.

In this context, health education implies specific programs of education directed toward people in the workplace. Its objective is to provide sufficient relevant information to help in the implementation of measures to prevent physical, psychological, or emotional injury or stress.

Initially, the emphasis of educational programs tended to be placed on teaching of lifting and handling techniques to workers, reflecting the high proportion of injuries that originated from this aspect of work. Later, classes were developed to prepare safety officers for a preventative role. At a later stage, it was realized that managers needed to be "sensitized" to ergonomics, so that they would ensure the establishment of suitable methods for the maintenance of the safety and health of their workers. Recent attention to the multiple causes of musculoskeletal injuries associated with repetitive tasks has highlighted the need to offer educational programs for

---

*Note:* For further information on this topic, refer to "Overview of Health Education; Uses and Limitations" by Margaret I. Bullock, appearing in *Health Education in the Workplace: Topics and Techniques,* May 13-15, 1987, a Pre-Congress Course, Australian Physiotherapy Association.

managers, supervisors, and workers so that each group can appreciate its own role in the prevention of injury or stress.

An overview of health education, from the point of view of the industrial health professional, helps to highlight the various processes involved in planning and implementing such programs and reveals both the uses and the limitations of health education as a preventive measure.

## PLANNING AN EDUCATIONAL PROGRAM

There are three important and closely related aspects of an educational program that need to be considered in planning: (1) the needs of the recipients, (2) the nature of the work and its inherent problems, and (3) the objectives of the educational program. Also to be considered are the content of the course, the effect of legislation, and the limitations of educational programs.

### Needs of the Recipients

In the working environment, many groups can benefit from a specific educational program. Industrial health professionals must consider their specific needs and design the educational program accordingly. The various groups could be classified as workers, supervisors, and managers or employers.

A variety of educational programs may be offered to workers, focusing on the specific needs of workers. For example, programs can educate

1. workers in general, who need to be given an appreciation of the principles of prevention and the methods of implementation
2. new employees, who not only need to understand ergonomic principles, but also may need to be introduced to certain work techniques and trained in appropriate approaches to movement
3. workers in specific areas, who need to be given an appreciation of the special demands of the work and introduced to methods of coping with them.

In addition, workers who have suffered previously from a work injury need to be guided in how (1) to avoid further injury, (2) to modify the workplace to ensure that it matches their current capacity or limitations, (3) to move their bodies in the safest way, (4) to modify their work movements to prevent a recurrence of the original injury, and (5) to apply preventive approaches in out-of-work hours. Other groups of workers also need special consideration. These include the aged, who need to be alerted to the likely changes in capacity and to the normal processes of aging, which in themselves present particular hazards (eg, the greater likelihood of hernias in older men), and the disabled, who need to be taught how to modify the task to cater to any specific incapacity, while maintaining a safe approach to work.

Safety officers and supervisors also need educational programs and need guidance not only in ergonomic principles of prevention but also in methods of teaching and supervising effectively.

Managers and employers require a slightly different approach in terms of education as they not only may need to be introduced to ergonomic approaches generally, but also might need to be convinced of the necessity to introduce an educational program to the workplace.

Although the basic principles offered to each of these groups may be similar, emphasis must be shifted subtly for each group to meet its special needs.

## Nature of the Work

In devising an educational program, the educator must appreciate not only the requirements of participants but also the nature of their work, so that the education is appropriate and meaningful. Initial assessment of specific problems during a visit to the worksite is essential for this purpose.

There are several categories of work that may be a source of potential problem to workers. Of particular importance are the areas of manual handling and lifting, repetitive activities (with or without precision), and sedentary work (with or without stress). Although they overlap, these broad areas give the industrial health professional the opportunity to consider particular aspects of a work problem that should be addressed.

### Manual Handling and Lifting

The worksite visit will reveal the special nature of the work and of the object to be lifted (eg, whether the object is animate or inanimate); weight, size, and bulk of the object; accessibility of the object to the lifter; space available for the lifter to grasp, lift, and carry the object; distance the object has to be carried; and frequency of lifting.

Although similar principles of lifting would probably be offered to all concerned with manual handling, different emphases would be placed in the educational program according to the tasks demanded of the worker. For example, slightly different points might be highlighted for nurses lifting in hospitals than for workers stacking boxes in a supermarket or for mechanics handling machinery in a workshop.

## Repetitive Work

Much of the work demanded in today's mechanized world is repetitive, and there are many settings where workers are involved in repetitive tasks. These activities may be defined as light tasks, consisting predominantly of physical work that does not require a great muscular force but that is characterized by high frequency of movement. The nature of the work can place different demands on the worker, depending on the requirements for precision, accuracy, or speed.

Psychological stress, which could be experienced by a typist working at a visual display terminal (VDT) or a worker assembling tiny units in an electronics factory, may add or contribute to the incidence of musculoskeletal distress after periods of repetitive activity. The demands imposed by "pacing" (ie, work rate set by the speed of the assembly line and the need for the worker to keep up with the line) can introduce similar stresses on the factory worker.

In all work situations involving a repetitive task, features, such as frequency of the task, magnitude of the load on musculature, duration of work without rest, range of joint motion, and force used, should be evaluated so that in the educational program proper emphasis can be placed on those particular areas needing special attention. By providing an understanding of the problems produced by some of these factors, the industrial health professional is likely to gain greater cooperation from the workers involved when changes are suggested.

There are many other features important to examine when planning the details of an educational program. For example, for workers concerned with either manual handling or repetitive tasks, consideration of the capacity of the worker for the task or the posture demanded by the task, may highlight the need for specific exercise or relaxation, or for modifying the design or layout of the work area, during the educational program.

## Sedentary Work

Although many people involved in repetitive tasks are seated, the category of sedentary work needs to be considered separately. The manager—coping with a multitude of tasks at his or her desk—air traffic controller—monitoring screens and making vital decisions for lengthy periods at a time—taxicab driver, and aircraft pilot could all be described as sedentary workers. Because they could suffer from physical, mental, or psychological stresses, they need to be educated in the hazards of sedentary work and methods of ensuring their continued well-being.

From the standpoint of physical stress the importance of workplace design and worker-task relationship needs to be emphasized. In such cases, education must be focused on design principles relating to the workplace and particularly to the design of a chair, seat, desk, or bench to match both the size of the sedentary worker and the particular demands of the work. Education would highlight the importance of the interrelationships involved, whether it be the manager seated at a desk and in easy reach of the telephone, the taxicab driver seated at the wheel of a car, or the aircraft pilot seated in the cockpit monitoring dials.

Fitness may prove to be a second area of physical need for sedentary workers. This could, of course, apply to workers in any category; but, for those who spend the major part of their working hours sitting down, the importance of education in fitness is especially great.

The emotional and psychological stresses imposed on some sedentary workers also suggest the need for education in stress management and relaxation. Education in this area may also be appropriate for workers in other areas (eg, those involved in highly repetitive tasks requiring precision or under paced conditions).

Regardless of the nature of the task—heavy or light, repetitive or intermittent, active or sedentary—personnel should discuss their perceived problems. Results of these discussions, together with the results of observations, would form the basis for planning the particular educational program required for specific participants.

## Objectives of the Educational Program

The third major area involved in planning an educational program is the definition of objectives. In part, this depends on the recipients of the program and the type of work in which they are involved.

Possible objectives for an educational program designed specifically for various categories of workers include:

- To introduce the concept of ergonomics.
- To provide an understanding of the mechanism of injury, relevant to the workplace.
- To explain the processes that may be used for injury prevention in the workplace.
- To encourage workers to question the suitability of their own environment.
- To teach new approaches to movement or new techniques of performing tasks.
- To demonstrate the need for physical fitness.

- To explain the worker's rights according to the relevant legislation or standards governing the workplace.
- To emphasize the worker's responsibility to follow the guidelines for safety provided at the worksite.

Considerable differences in approach are needed for educational programs for supervisors. The objectives that apply to programs for workers may well apply to supervisors. However, in addition, because safety officers and supervisors will need to offer "in-house" courses to workers, objectives for their courses could also include:

- To provide a reasonable depth of understanding of body mechanics, mechanisms of injury, design principles, and work techniques.
- To provide an appreciation for the particular demands of work in a variety of settings.
- To demonstrate methods of teaching that would be appealing to the workers.
- To outline approaches to supervision that would ensure cooperation from workers.
- To introduce efficient methods of data collection relating to incidence of injury.
- To ensure appreciation of legislation and standards relating to the workplace.
- To emphasize the responsibilities of supervisors within the work establishment.

Quite different approaches may be needed in the education of managers. For example, the objectives may be:

- To sensitize the manager to the need for ergonomic programs in general.
- To convince the manager of the need for an ergonomic prevention program in his or her work area in particular.

Further objectives of an educational program for managers could include:

- To provide an overview of relevant legislation and standards.
- To ensure an understanding of managers' responsibilities for the safety and well-being of employees.

When managers have identified particular problems in their own work area and perceive the need for specific assistance, the objectives may also be:

- To educate in ergonomic principles of prevention.
- To provide guidelines for reduction of stress.
- To improve physical fitness.

**Content of the Course**

The content of the educational program depends on the participants, their workplace, and the stated objectives for the course. A number of components can be included in health education programs. For example:

- The importance of ergonomic prevention programs can be appreciated if participants understand the way the body moves and works; and, for this reason, some basic instruction in relevant anatomy should be included to provide a basis for the introduction of simple concepts of the biomechanics of movement. The principles of good static and dynamic posture establish a firm foundation for later instruction on appropriate movement methods or on positions assumed during work.
- The concept of balance and stability is important for those involved in manual handling and lifting, and the educational program should demonstrate the effect on the musculoskeletal system and its stability of having a narrow base of support, a high center of gravity, or a load positioned too far from the line of gravity.
- Knowledge of the limitations imposed by anatomical structures on body movements or positioning can help participants to appreciate why injuries can occur in certain circumstances. This can provide insight into the need for application of certain guidelines.

One of the most important aspects of prevention is the design or redesign of the workplace and the implementation of correct working methods. Educational programs must highlight the design problems that the ergonomist perceives within the work station. With this background, participants are able to view films of their own workplace and identify potentially hazardous features of design or body movements. All participants must be able to relate ergonomic principles to their own working environment and to discuss their own problems.

Physical and psychological stresses at work is not a new phenomenon, and education in methods of coping with these stresses is essential for the physical and mental health of those involved. The importance of appropriate and timely exercise and relaxation as well as general physical fitness needs to be emphasized; and the possibilities of job rotation, exercise breaks, physical fitness classes, and stress management sessions need to be considered, as appropriate.

As the majority of work-related injuries are associated with lifting and handling, due emphasis must be given to this type of education when lifting and handling form part of the work process. However, problems may emanate from some other cause; and, in such cases, lifting and handling education must be supplemented with a program directed more toward the particular work demands. The change in approach to work and the increasing use of computerized equipment has led to an increasing proportion of people who carry out their daily activity in the seated position. Education in correct sitting must form part of the emphasis on design of the workplace; but it also must be related to anatomy, posture, stability, movement dynamics, and the mechanism of injury. Educational programs focusing on seating requirements should emphasize the variability of body size and the importance of recognizing individual worker body dimensions when defining the chair, desk, table, and bench relationships for the particular worker.

When managers are being introduced to the value of ergonomics, an explanation of the problems that can occur through inadequate workplace design and the methods that can be used to prevent these problems may be sufficient to present a global view of ergonomics. When the objective is to convince the manager that his or her company needs some ergonomic intervention, then more specific techniques of "persuasion" would need to be included in the educational program. For example, data relating to injury frequency, costs to the company in lost time or compensation payments, comparative figures in other companies that have adopted ergonomic principles, cost of introducing preventive measures, likely benefit to employees of introducing ergonomic principles, possible increase in productivity, and likely cost benefit to the company are some of the items that might need to be included in an educational program.

**Effect of Legislation**

When planning an educational program it is important to be aware of the nature of the relevant legislation, standards, and union practices that apply to the particular work area. There is little point in advocating certain practices if laws and industrial agreements prevent their implementation. It is also essential that all categories of participants be made aware not only of their rights but also of their responsibilities according to the law.

## Limitations of Educational Programs

Ideally, educational programs will be effective in changing attitudes and work practices. However, few working places are "ideal"; and, despite the best effort of industrial professionals, educational programs can have their limitations.

It is presumed that better informed workers will apply safety principles more willingly and will modify their work methods more effectively, particularly if they have been involved in defining solutions. However, there are a number of reasons why the health education program may not be as effective as was hoped. These may lie within the program, or be associated with the worker, the supervisor, or the employer.

In planning the health education program, a certain level of basic knowledge in the participant must be presumed. If this presumption is incorrect, then the educational program may be pitched at too high a level for the workers, or it may move at too great a pace for their total comprehension.

The time available for the program also is an important consideration. The optimal duration both for one session and for the total program should be determined. As the amount of material absorbed in any one session is limited, the value of continuing past this point is dubious. The interest of participants varies; and, if not well motivated to learn, they may absorb little during the teaching sessions. It is a challenge to the educator to capture and maintain the interest of participants by providing varied activities, by encouraging active participation, and by referring frequently to relevant features of the participants' workplace. The participant must be so convinced of the value of the advice that he or she is committed to applying new work methods or accepting a redesigned workplace. Without this commitment, there may be great difficulty in implementing new work methods.

Competing interests may interfere with the workers' motivation to apply ergonomic principles, despite their introduction to the hazards of certain activities. For example, job rotation may be suggested to workers who are paid on a piecework basis. Their interest is in speed of activity and in producing many items in one day. To rotate to an unfamiliar job in which they are less skilled will almost surely mean decreased output and payment. The workers' incentive is to continue in the old, known method and not to incorporate changes that might interfere with speed and production rate.

Although training and instruction might be given in new and more appropriate methods of working, the responsibility lies with the worker to comply with these new instructions; and the effectiveness of the educational program can be reduced if such responsibility is not accepted.

The amount of information retained by the participant after an educational program will vary. For those not accustomed to listening to lectures or attending classes, attention level may be quite low; and the value of the program will be limited if care has not been taken to provide clear examples and relevant experience that would help to capture attention. There is no guarantee that the participant will retain the newly acquired knowledge unless appropriate reinforcement is offered at a later date. As a supplement to classes, handouts in terms the participant could easily understand can be distributed. Follow-up programs could also be offered at appropriate intervals to reinforce lessons already given and to enlarge participants' knowledge.

The effectiveness of an educational program also depends on the availability of someone to whom the participants may refer for ongoing advice. Despite good intentions about applying safety principles, workers may have difficulty remembering details and their approach may need review, modification, or correction. The importance of having a supervisor at the workplace cannot be overemphasized. An experienced safety officer may fulfill this role, but the person responsible for offering the program may also prove to be the most effective advisor.

Supervisors can be very useful in the implementation of these programs. However, their level of knowledge and motivation are important for success. The value of the program will be limited if supervisors are not committed to assessing problems or to discussing possible solutions with employees. Their reporting of accidents and identification of causes is important for the design of further educational programs and of corrective measures. Without such input, the industrial educator has more difficulty in effecting improvements to an educational program.

The attitude and commitment of the employer are also critical to the long-term success of health education programs. Employers' positive attitudes can engender cooperation, interest, and enthusiasm in employees. The sensitization and education of employers in the importance of ergonomic preventative measures is essential, and sufficient time must be devoted to this task for successful implementation of principles advocated at different levels.

## CONCLUSION

Health education in the workplace is an important component of the overall ergonomic approach to injury prevention. Careful planning to ensure relevance to the participants can help to improve the effectiveness of educational programs in work settings.

# Section 3

# Back Schools

*Susan J. Isernhagen*

## INTRODUCTION

Back schools in the United States are a relatively new phenomenon. They began to gather strength in the late 1970s and early 1980s. The concept is so logical and the results have shown such potential that back schools are now becoming an important part of prevention of injury in the workplace. The following are excerpts from several "experts" in the back school and injury prevention profession. They participated in this forum by answering questions in letter format. Their responses were directed to the subjects of back school effectiveness, role of fitness instruction and education in prevention programs, ideal injury prevention programs, and thoughts for future direction.

## ARE BACK SCHOOLS EFFECTIVE IN PREVENTING INJURY?

**David Apts, PT:**
*American Back School*
*Ashville, KY*

Yes, back schools are effective in preventing back injury. One example collected from the industry highlights the experience of Westmoreland Coal Company. In 1983 I was hired to do a three-hour program on preventing back injury as part of the annual retraining. The number of participants in the program varied from 10 to 35 on all shifts. Over 1,500 employees were trained between January and October. No follow-up training was scheduled in 1984. The injury reduction rate between 1983 and 1984 was 330 lost-time accidents. In calculating the cost savings, employees' salary, benefits, and productivity were measured. Total savings in one year was $161,130.

**Mary A. Mistal, PT:**
*Back Education Consultant*
*Billings, MT*

In 1983 at the National Safety Council Conference, I heard Bill Mattmiller, PT, founder of one of the first back schools, state that back schools were ineffective in preventing back injuries in industry. After this devastating statement, Mattmiller went on to emphasize that unless management became an integral part of back injury prevention programs in industry, the program would fail. After hearing this, I revised my back injury prevention programs to emphasize upper- and mid-management involvement.

I introduced my initial program to an airline based in Hawaii. The focus was basically on back injury prevention, incorporating basic anatomy of the spine, intradiscal pressure studies, posture, positioning, and body mechanics. The program was specific to the job-related activities of the airline using its own employees to perform the job tasks. A slide-tape presentation was prepared using these actual situations. Live demonstrations were then scheduled to ensure that all employees participated.

Upper- and mid-management were represented at each of the subsequent live demonstrations when further discussion occurred regarding specific problems the employees faced. This dialogue helped not only to eradicate morale problems but also to resolve ergonomic and equipment problems encountered with repetitive sporadic lifting. To further assist this effort, a back booklet specific to this company was written and published.

Three years later, this company had reduced its losses significantly by reversing the loss in 1983 to a gain of $500,000 in the first quarter of 1986. New employees and furloughed employees have been added to the work force, and expansion of services has taken place. A complete reversal of the devastation facing this company has resulted, partially due to the back injury prevention and wellness programs and, most importantly, to the strong commitment of upper- and mid-management to perpetuate these programs.

## Ronald W. Porter, PT:
*Back School of Atlanta*
*Atlanta, GA*

Back schools do not prevent all back injuries, but they appear to be helpful in decreasing the duration and severity of back problems and in returning people to work earlier. This is supported by decreases in workers' compensation claims by facilities at which our back school has helped conduct back school programs. In the Chelsea Back Program, after implementing a back school, its incidence rate of back injuries increased from 2 to approximately 9 per 200,000 manhours. However, the number of lost-time injuries decreased, and the workers' compensation claim expenses decreased from $218,000 in 1979 to $18,000 in 1982.[1]

## Nancy C. Selby, BS Ed:
*The Spine Education Center*
*Dallas, TX*

In 1982, the Spine Education Center in Dallas asked an independent research team to conduct a study of eight industrial back school programs that had been customized for specific industries and had been used by the companies for a year or more. The average reduction in back injuries was 25% and the average reduction in lost-time days was 50%. Since that time, we have requested that every client keep statistics so that we can evaluate the effectiveness of the back school programs. The statistics have remained constant since that time if a program was used the way it was designed.

## Arthur H. White, MD:
*San Francisco Spine Institute*
*Daly City, CA*

From our personal experience in San Francisco we have helped at least 100 industries set up educational and exercise programs and have followed their statistics. The only one of these follow-up studies that has been published is the Southern Pacific study that has been coordinated by Bill Mattmiller. The current statistics are not available to the public. The previous statistics, although not on a controlled study, showed a dramatic decrease in cost and injury rate. We have had similar experiences with Safeway, which does collect its statistics (available, but not published). The simple fact that these organizations continue to spend considerable dollars on education and exercise while keeping their own statistics is certainly evidence that it is working for them.

There are many complexities that prevent adequate studies from being performed. The psychological and socioeconomic factors alone are enough to invalidate almost any study. Back pain is also such a vague, long-range, and aging process that we are dealing with apples and oranges. I also have yet to see a controlled, double-blinded study that demonstrates that tooth brushing or fluoride reduces tooth decay. That may seem trite; but, with all the personal success and basic science data to draw from, it seems unwise to quit doing what is working well while we wait for science to catch up.

## WHAT IS THE ROLE OF FITNESS INSTRUCTION AND EDUCATION IN PREVENTION PROGRAMS? IS IT HELPFUL?

**David Apts:** Yes.

**Mary A. Mistal:** Yes.

**Ronald W. Porter:** Yes.

**Duane Saunders, MS, PT:**
*Educational Opportunities*
*Edina, MN*

There are two areas in which we should be interested. One is the prevention of musculoskeletal injuries, which would involve exercises to strengthen and improve flexibility specific to certain job requirements. The other is cardiovascular fitness and the concept of aerobic fitness training. Both are important and would be beneficial for companies to initiate. However, we need more substantiation and proof that these programs are cost effective.

**Nancy C. Selby:**

If fitness is to be incorporated into the workday, management would accept a program that requires a minimum of time and equipment. Some simple stretching (ie, the kind an athlete would do before an event) and strengthening before work commences might be very effective. Both flexion and extension stretching should be used. Abdominal muscle and quadriceps strengthening would be beneficial as long as a program could be done without extra equipment. There are companies that are very proud of the fact that health club memberships are paid for their employees. Unfortunately, management is more prone to use this than the blue collar employees who are having the significant back injuries. (See Pause Gymnastics—Exercise Breaks in Section 4-2, *infra.*)

## WHAT WOULD THE IDEAL INJURY PREVENTION PROGRAM CONTAIN?

**David Apts:**

The injury prevention program that would most benefit industry would (1) show people what their bodies are like, using anatomical and physiological slides that are straightforward and not complex; and (2) show people, as they work at their particular work spaces daily, how they work, how they can improve their habits and, also, the ergonomic components that need to take place to change job sites.

A large dose is necessary at first; therefore, a three-hour program for labor-intensive employees is best. After that, weekly safety talks given by the workers' supervisor are important, even if they last only five minutes. Poster campaigns of pictures at the worksites can be blown up and used on a weekly basis. Some type of one-on-one follow-up by supervisors also is effective. The employee can then be approached on the job and helped in whatever way needed. With respect to follow-up, every 1 to 2 months some type of effort should be put forth regarding the proper methods of bending, lifting, pushing, pulling, and exercise.

**Mary A. Mistal:** (See answer to first question.)

**Ronald W. Porter:**

The ideal industrial injury prevention school should have classroom education as well as individual sessions on problem areas and a self-study component.

Classroom work can be taught to groups of approximately 25 people by an accredited professional on the topic. In a back school a physical therapist or a physician is the best choice. Instructors should cover information on the structure of the spine, what can go wrong, and how to prevent the problem from occurring. Audiovisual materials are helpful as are visual models and posters and workbooks, if the participants can or will read. These sessions should be conducted annually with 6-month follow-ups.

Individual sessions can be conducted one on one by the instructor with the employees that have demonstrated particular need in the instructor's areas of expertise. During these sessions demonstrations and practice of prevention or helpful technique is important. Videotaping of these sessions can be helpful for review by both the instructor and the participant.

Self-study by the person can be accomplished at work or at home using videotapes and workbooks, when either the employee or the employer deem it necessary.

**Duane Saunders:**

The ideal industrial injury prevention school, including back school, should have the following components:

1. Worksite evaluation and recommendations to change the worksite. Many of these recommendations come from the workers themselves. Therefore, the idea of worksite modification and evaluation of the workplace should be incorporated into the educational program to get supervisors and workers to think about this area.
2. Discussion about the myths and misconceptions as far as back care and back injury are concerned. For example, the idea that back injuries are the accumulative effect of what we do rather than a single event injury should be stressed.
3. Review of all the ways the back can be harmed, such as through poor posture, faulty body mechanics, and stressful working and living activities; development of stiffness in our backs; and lack of physical fitness. These seem to be the real causes of back injury and should be pointed out.
4. Review of some of the types of back problems, such as muscle spasm, disc injuries, arthritis, strains and sprains, and so forth.
5. Emphasis of the correct ways of doing things—in other words, good body mechanics and healthy ways to do a job.
6. Program closure with exercise classes to determine the level of flexibility and strength of participants to show them exercises to improve strength and flexibility.

An initial three-hour program for groups of 30 participants or less is preferred. This includes 30 minutes of actual practice with exercises and body mechanics. Program follow-up should be at 3 to 6 months, with a shorter (half hour), one-to-one review session. An expert physical therapist should do the presentation. Training of supervisory personnel as an alternative has not been totally effective.

**Nancy C. Selby:**

The ideal industrial injury prevention program should emphasize body mechanics modifications both on and off the job and the benefits of physical conditioning. The program should be short enough to keep the attention of the blue-collar worker and should be relevant to his or her job situation. Time is money where management and production are concerned, so the time spent on an injury prevention program needs to justify the means. Anatomy of the spine, first-aid for back pain, stress, and nutrition may all be worthwhile components of an injury program; but they are secondary to the body mechanics training.

Employees should also participate in body mechanics activities that are relevant to their job. They should practice lifting the material they handle in their job. The same should occur if they have to push, pull, pivot, stand, or sit. They need to know how to modify their activities without decreasing production and to increase their level of awareness so they can better protect themselves. Understanding the transfer of these techniques to other jobs or home situations is equally important.

Small classes will be more effective for high retention levels, and the programs should be repeated at least once a year (every 8 or 9 months would be more beneficial). Follow-up materials also are essential. The program should be designed initially for people who are hurting themselves but also should be given to everyone in the industry because of the prevalence of back injuries in our society.

The trainer or teacher is the motivator and will be responsible for compliance with the back school concepts. Historically, supervisors have been responsible for accident prevention in industrial settings. However, few of these people are really trained to disseminate the information effectively. They are more effective at reinforcing rather than teaching the basic concepts. Usually an outside trainer, who is very familiar with the information as well as the industrial situation, will be the most effective teacher. If that is not feasible from a cost standpoint, a mid-management person can be trained to give the information to the blue-collar employees.

**Arthur H. White:**

For the most effective prevention program employees need to be motivated at their level of understanding by someone they respect. Other recommendations include:

- Find motivating leaders of the sector that the program is geared to help and train them to be trainers.
- Give the participants simple messages.
- Require that the employee participate and be tested on his or her understanding and accomplishments; reward the employee for this.
- Demand and reward a general exercise program with an emphasis on the muscles that can potentially decrease back injuries.
- Make the program interesting and make it live.
- Use trainers who know how to gain audience participation.
- Do not use audiovisuals or dry lecture format.
- Do as Safeway has done and have wall slide contests with a vacation for the winner.
- Use all the advertising gimmicks used by Wall Street to subliminally influence the employees.

## WHAT ARE SOME CHALLENGES FOR THE FUTURE IN INJURY PREVENTION PROGRAMS?

**David Apts:**

Our future challenges are:

1. We (in the profession) need to know how to provide better back school education and to understand better the components of that education.
2. We should have standardization within the dissemination of the back school information.
3. We need to find the ideal program format and to be able to show beyond a shadow of a doubt, to industry, the cost effectiveness for the return on the investment. Incorporation of back school education, ergonomics, and fitness into the workplace is the only way to combat back injuries.
4. We need to develop better treatment alternatives that are conservative, not surgical. Physical therapists are at the forefront of doing this.

**Mary A. Mistal:**

The challenge to industry is whether or not to make the commitment to initiate and promote an ongoing awareness of back injury prevention and wellness programs among all of its employees.

**Ronald W. Porter:**

The challenge for the future is to convince industry and the general public that health education and prevention are viable health management strategies. For this we need documented evidence and vocal health professional advocates.

**Duane Saunders:**

We need qualified instructors for our injury prevention programs. Unfortunately, many people, including physical therapists, are getting "on the bandwagon" to provide these types of programs. In some cases hospital management and people in the health care field see this as an opportunity to make money and get involved in these programs with very little experience and expertise. This is unfortunate. Just because one is a physician or a chiropractor or a physical therapist does not necessarily mean that one is an expert in giving this kind of educational program. In fact, there is no guarantee that the person has the teaching skills that are necessary, even if he or she knows the material well. If everyone regardless of skill and background attempts to do this, there will be many failures and many companies will be turned off because of the poor quality of what is being presented.

Our goal is that only those people who are well qualified and who are good teachers will attempt to present these programs. If this is the case, we will have significant success in this area.

**Nancy C. Selby:**

As the costs of medical and legal fees continue to rise and more companies are finding insurance costs prohibitive, prevention programs and self-help education programs will increase. It is one of the few avenues available to industry. At the same time, industry will have to recognize the essential component of good ergonomic conditions, the purchase of mechanical devices, and the advent of robotics to assist employees. Employees are also going to have to take some responsibility for their own health care.

We have just touched the tip of the iceberg in the area of back injury prevention and health care responsibility. Back education programs might even become a mandatory part of every industrial orientation program.

**Arthur H. White:**

There is definitely an awareness and forward movement of these types of industrial programs. Ten years ago there were only one or two well-known industrial programs in the world; now there are thousands. It requires a corporate commitment. When we were dealing with middle management without the support of dollars from the top, we were working with our hands tied. Now that most large industries are self-insured, they can better see their losses and the benefits of these programs.

Originally, these programs were conducted by a physical therapist who did not have specific communications background and skill. Now people with a background in communications or psychology can put together packages that are much more palatable and therefore potentially more successful. The audiovisuals, of course, have greatly improved. The statistics are mounting. More grants are being issued for this type of study.

We are gaining momentum. This momentum will mount over the next 5 years to reach an absolute flurry of new programs in industry and public schools. We will see public television commercials. We are already seeing back articles in every popular magazine. In 10 years there will be back programs virtually in every major industry, and in 20 years back care will be as common as hard hats, protective eye goggles, handicapped parking, wheelchair curbs, and the use of seatbelts. None of those things were in common use 10 years ago!

---

**REFERENCE**

1. Fitzler SL, Berger RA: Chelsea Back Program: One year later. *Occup Health Saf* 1983; 52:52–54.

# Section 4

# Industrial Health Programs

# 1. Fitness and Health Promotion Programs in Industry

*Susan J. Isernhagen*

## INTRODUCTION

Company sponsorship of health programs within business and industry is a response to the awareness that healthy people are also healthy, productive employees. This is evident in the companies in the general well-being of the work force, which influences attitude; the measurement of productivity, which is higher with a healthy work force; and the prevention of injury or other disease, which may affect productivity and workers' compensation or health insurance claims.

Andrew Wood, RPT and exercise physiologist, is the supervisor of Health Promotion and Fitness at General Mills, Inc, Minneapolis, Minnesota. He has instituted a considerable number of health promotion programs within that company. Wood explained these programs in personal correspondence to the author and in "The Basics of Worksite Health Promotion."[1]

## FITNESS WORKS

Fifty percent of our health is determined by our lifestyle; that is, our diet, whether or not we smoke, level of activity and so forth. Consequently, many public health organizations and businesses have started health promotion programs in an attempt to make people aware of how life style affects their health, to increase productivity, and to decrease health care costs. There are a variety of ways to promote health in the worksite, from large elaborate fitness centers to self-directed, healthy life style programs.

There are many subjective and objective reports that show the benefits of good employee fitness to industry. These benefits include a decrease in the number of cardiac risk factors (eg, lowered cholesterol level and lowered blood pressure) and an increase in the level of fitness (ie, an increase in maximum oxygen consumption) and flexibility.[2,3]

Bly and associates[4] reported on a comprehensive health promotion and health screening program at Johnson and Johnson. The participants in the program experienced greater levels of fitness and a decrease in coronary heart disease risk factors along with a decrease in inpatient health care costs as compared to those who did not participate in the program. Wood and colleagues,[5] at General Mills, reported a decrease in cardiac risk factors along with a 19% decrease in absenteeism for employees who participated in a health awareness program for one year.

Fitness, however, is only one component of a health, safety, and wellness program. Ideally, the program should incorporate smoking cessation, stress management, chemical abuse counseling, blood pressure management, nutrition education, and recreation along with physical fitness. All participants should take a health risk appraisal to determine what life style health risks they have and need to change to improve their health. The participants also should have a physical examination. This should include a stress test—to determine maximum oxygen consumption—and tests for flexibility, strength, and body fat composition.

## PHYSICAL FITNESS

Fitness classes should be tailored to the level of fitness of the participant. Physical fitness can be determined from a cardiac stress test and a history of the person's on-the-job and off-the-job activities. A qualified fitness instructor is a must. The instructor should be certified in cardiopulmonary resuscitation; have a good knowledge of exercise, anatomy, and injury prevention; and be certified as an instructor with the American College of Sports Medicine or similar professional organization.

Each exercise session should include three phases: warm-up, aerobic exercise, and cool-down.

The warm-up phase comprises flexibility exercises and slow, rhythmic movements to prepare the body for the upcoming activity.

The aerobic exercise phase should be 20 to 30 minutes of rhythmic activity with an intensity between 60% and 80% of the participant's maximum aerobic capacity as determined by the cardiac stress test. Activities that increase the demand for oxygen include walking, cycling, swimming, cross-country running, skiing, and aerobic dance. Aerobic workouts should be done three to five times a week. Weight training can also be incorporated into the exercise phase of the workout; however, weight training is not aerobic and cardiovascular benefits will not be seen. Weight training can include endurance programs involving low resistance with many (10 to 20) repetitions. Strength training involves high resistance with few (5 to 10) repetitions.

The last phase, the cool-down, involves the activities done in the aerobic exercise phase, but at a slower pace, and mild stretching. This prevents blood pooling and orthostatic hypotension.

Workouts should also encompass exercise activities for the person's specific on-the-job or off-the-job activities. This gives the employee an excellent opportunity to practice proper lifting and body mechanics with the exact weight or object he or she is supposed to lift on the job or at home.

## SMOKING CESSATION

All health promotion programs should have a smoking cessation component. Smoking has been recognized as the largest cause of preventable deaths in the world. Several studies have shown that smokers can cost a company between $600 and $4,000 more per year in health care costs than nonsmokers. There are many "quit smoking" programs available. The "magic pill," such as hypnosis and acupuncture, can be initiated successfully; however, the success rate of these methods after one year is poor. Scare techniques have also proved to be unsuccessful. Methods that use behavior modification and group support are usually the most successful. Smoking is an addiction; therefore, the most reasonable success rate after one year is about 25%. Smoking cessation classes should meet once or twice a week for several weeks and address such topics as ambivalence, mastering the social obstacles of quitting and restarting, stress management, exercise, weight control, and body function. Follow-up after the classes are completed is essential.

The most effective smoking cessation activity is to implement a no smoking policy. The main rationale is to limit the exposure of second-hand smoke to the nonsmoker. Companies like Honeywell, General Mills, and Multimedia all have restricted smoking policies.

## HYPERTENSION

Hypertension is one of the major risk factors for cardiovascular disease. Employee education of hypertension can be done through classes, displays, and blood pressure screenings. The main objectives of a blood pressure management program include:

- detection and referral for proper medical care;
- evaluation of the general health of the hypertensive employee; and
- follow-up care with the employee's personal physician.

## STRESS MANAGEMENT

Stress management is becoming more important in the worksite health promotion program. Self-relaxation can not only relieve the day-to-day tension but also lower blood pressure. Support groups can be formed to deal with topics such as self-relaxation techniques, goal setting, and positive thinking. One-on-one consultation with a stress management counselor also can be beneficial.

## WEIGHT CONTROL

Weight control is always a very popular subject and most programs will be well attended. Like any popular program, there are a variety of methods to choose. One of the best set of guidelines to use when considering a weight control program is that of the American College of Sports Medicine. These guidelines state that a proper weight control program should include:

- no more than a 1.5 to 2.0 lb weight loss per week;
- a caloric intake of no less than 1,200 calories per day;
- an exercise component, with the person working out three to five times per week at 60% to 80% of his or her predictive maximal aerobic effort;
- a diet that contains a variety of foods; and
- high fluid intake.

## SUMMARY

Table 4.1-1 provides a list of various fitness and health promotion programs, including topics, frequency of the classes, and recommended instructor.

**Table 4.1-1** Fitness and Health Promotion Programs

| Class Topic | Frequency | Instructor |
|---|---|---|
| Safety programs | Once a week | Safety representative |
| Back school | Once a week for 3 weeks | Physical therapist |
| Back fitness | Three times a week for 8 weeks | Physical therapist |
| Work hardening | Eight hours a day for 6 weeks | Physical therapist or occupational therapist |
| General fitness | Three times a week | Physical therapist, nurse, or certified fitness instructor |
| Smoking cessation | Two times a week for 4 weeks | Physical therapist, occupational therapist, psychologist, nurse, or physician |
| Nutrition education | Once a week for 10 weeks | Dietitian or nutritionist |

Within industry, the greatest challenge for the future in the health and wellness area is to affect the blue-collar worker. Wellness programs have been successful in the white-collar work force, but these programs have not caught on well in the blue-collar sector. Blue-collar wellness is a prime goal.

### REFERENCES

1. Wood A: The basics of worksite health promotion. *Network* 1986, 3(3):1.
2. Blair SN, Piserchin MS, Willar CS, et al; A public health intervention model for worksite health prevention. *JAMA* 1986;255:921–926.
3. Hannan EL, Graham JK: A cost benefit study of a hypertension screening and treatment program at the work setting. *Inquiry* 19;15:345–358.
4. Bly JL, Jones RC, Richardson JE: Impact of worksite health promotion on health care costs and utilization. *JAMA* 1986;256:3235–3240.
5. Wood EA, Collins JJ, Halaney ME, et al: An evaluation of the TriHealth Program after 1 year at General Mills, Inc., in Opatz JP (ed): *Health Promotion Evaluation: Measuring the Organizational Impact*. National Wellness Institute, University of Wisconsin—Stevens Point, 1987.

# 2. Pause Gymnastics—Exercise Breaks

*Susan J. Isernhagen*

Anyone who works in a static position or with repetitious dynamic movements recognizes the comfort benefits of changing position or activities. Although many people do this automatically, the pressures and payment schedules of today encourage people, at times, to work beyond their comfort level. Exercise breaks or pause gymnastics are designed to reduce the physiological effects of cumulative trauma.

The author gained specific information on the development of a pause gymnastics program in personal correspondence with Amanda Gore, PT (on August 7, 1987), and from the published text by Gore and Tasker.[1] The following are excerpts of the correspondence in which Gore discussed factors to be considered in implementing a successful program.

## PRACTICAL ASPECTS OF THE PROGRAM

Once an employer in industry has decided to implement a pause gymnastics program, its success is dependent on the way it is "sold" to the workers and then on the way it is monitored.

It is important to educate workers about the benefits of, and reasons for introducing, the program and to assure them it will be fun *and* healthy. Introducing the exercises in this way and allowing the workers to feel a sense of involvement in the program, by encouraging their comments on the choice of exercises, reduces the chance of resistance.

The first stages are important to the continued success of the program. Levels of motivation must be maintained and the task usually falls on the manager, supervisor, and any enthusiastic worker.

Boredom can be avoided by offering the participants two or three exercise routines initially and then allowing them to select new exercises from a reference.

A successful method to use in introducing a pause gymnastics program follows:

1. Interview the manager and determine what time workers are available for participation. Can he or she afford to allot 3 five-minute breaks per day (the optimum)? or only 2?
   Exercise breaks are designed to allow recovery from fatigue, to prevent the development of aches and pains, to wake up the participants, to allow them to have some fun, and to encourage a feeling of well-being. Ideal times are mid-morning, after lunch, and mid-afternoon. Five or ten minutes of exercise is preferred; however, any exercise time is better than sitting or standing all day.
2. Arrange for a time to meet with workers, introduce the concept, and request their ideas and feelings. Arouse their interest and assure them it will be enjoyable and beneficial.
3. Create programs with exercises based on an investigation of the postures involved in the jobs of the participants. Finish each program with a dynamic activity to increase circulation and to "wake up" the workers.
4. Elect a "captain." This person is responsible for monitoring the effectiveness of the program and worker attitudes and highlighting any problems that may need to be rectified. Two or three "subcaptains" should be elected to assist the captain and to help lead exercise sessions. To ensure maximal participation, draw up a roster and assign each worker to lead the session on specified days. The worker and the subcaptain lead together when possible.
5. Introduce the first program and carefully explain the value and purpose of each exercise. Supervise the practice of the exercises and then allow the participants to use the exercises for a week. Explain you will return in one week with the next program, and, again, observe their practice of the new set of exercises to ensure they are performing them correctly. Continue on a weekly basis until they have all the programs. Then, they can mix and match when necessary and create new programs.
6. Be very careful to provide written instructions on how the exercises are to be performed.
7. Spend extra time with the captains and subcaptains to ensure that they really understand how to do the exercises properly and how to promote safe exercise techniques by being able to recognize unsafe movement patterns workers may adopt.
8. Provide to leaders, workers, and managers written information on how to perform the exercises and on their responsibilities.

9. Monitor the program at least weekly initially to prevent problems from becoming major—your involvement at this time is critical for success.
10. Provide new exercises if interest is waning. Ensure all workers who are experiencing difficulties that they have access to a health professional with whom to discuss their problems.
11. Have any workers who have preexisting medical or pain conditions examined by a medical practitioner before commencing the exercises.

Problems often arise if workers are self-conscious; it is important to request that all workers in the area during exercise breaks either participate or leave. Flexibility and a sense of humor greatly assist in the success of these programs.

Implementation also has some basic premises (described in Gore and Tasker[1]). Five golden rules should be explained to the workers to understand their own body:

1. Keep your body balanced.
2. Listen to your body and act on its warning signals.
3. If you work with joints in one position, change that position frequently (every 15 or 20 minutes).
4. If you work muscles statically, move them.
5. If you work muscles dynamically, relax and stretch them.

Stress management can also be accomplished with the assistance of pause gymnastics. Relaxation is as important as strengthening and stretching. Some of the increased physical signs that pause gymnastics can affect are those related to the "fight or flight" physical reaction to stress. The exercises would facilitate relaxation of muscles, a decrease in heart and respiratory, rates and, potentially, a decrease in blood pressure through relaxation.

## BENEFITS OF THE PROGRAM

The education received in the exercise sessions can affect total body awareness. The worker will be able to feel tension, stretching, relaxation, and the precursors of soreness. This is helpful on the job to prevent cumulative trauma and off the job to make appropriate life style modifications. Early recognition of muscle problems and effective intervention mean health for the worker's total body.

Therefore, pause gymnastics principles are not only health promoting, but preventative (of muscle injury). Although benefits are realized in the musculoskeletal system first, they can be viewed from a holistic approach as they are eventually realized in other systems of the person involved.

---

**REFERENCE**

1. Gore A, Tasker D: *Pause gymnastics—Improving comfort and health at work*. North Rhyde, NSW, Australia, CCH Australia Ltd, 1986.

# Section 5

# The NIOSH Strategy for Reducing Musculoskeletal Injuries[1]

*Timothy J. Pizatella, Roger M. Nelson, David E. Nestor, and Roger C. Jensen*

## SCOPE OF THE FIELD

Prevention of occupational injuries is one of 15 prioritized areas identified for improvement by the U.S. Surgeon General.[1,2] Among the objectives outlined are reducing the rates of workplace facilities, disabling injuries, and lost workdays due to injuries. In line with meeting these objectives, the National Institute for Occupational Safety and Health (NIOSH), the government agency responsible for directing the nation's occupational disease and injury prevention research programs, identified the ten leading work-related disorders. Musculoskeletal disorders resulting from exposure to manual load handling, repetitive motion, or vibration are listed as the number two research priority by NIOSH, second only to occupational lung diseases.[3]

Musculoskeletal disorders as a group constitute the largest percentage of work-related injuries in the United States. Low back injuries are the most frequent and costly of the musculoskeletal disorders. Musculoskeletal injuries to the trunk account for approximately 32% of all compensable injuries and 42% of the compensation costs.[4] Labor-intensive industries experience the largest incidence rates of compensation claims for back injuries; the mining and construction industries average more than 1.5 claims per 100 employees per year.[5]

In addition, the financial loss to the U.S. economy is staggering. Back injuries alone cost the nation an estimated $14 billion per year.[6] While the

---

[1]This chapter was originally written as a paper for the National Institute for Occupational Safety and Health, Division of Safety Research, Morgantown, West Virginia, and is considered a United States government work for which no copyright can be held.

pain and suffering cannot be measured, they too have a significant impact on the social and psychological well-being of the injured worker.

The NIOSH goal has been one of preventing musculoskeletal injury. The primary approach has focused on task evaluation and tool and workstation design. A secondary effort has been to improve clinical methods for objectively determining signs and symptoms associated with low back complaints. With the knowledge gained from improved injury definition, the ultimate strategy designed to prevent the injury through improved workplace design may be approached. To facilitate achievement of this goal, NIOSH has developed a proposed national strategy for the prevention of musculoskeletal disorders.

## THE NIOSH STRATEGY

The NIOSH strategy is based on the premise that the risk of individual workers experiencing a musculoskeletal problem is affected by personal factors and job factors. Thus, the goal of reducing risk requires programs that address both personal and job factors. To effect this approach, the NIOSH strategy includes a list of areas that need to be addressed if a significant reduction in the number of musculoskeletal injuries is to be realized. These areas are presented in the following excerpt from the NIOSH *Proposed National Strategy for the Prevention of Musculoskeletal Injuries.*[7]

### Refining Surveillance Systems

*Health Surveillance*

Strategies designed to prevent or mitigate musculoskeletal injuries require sensitive and verifiable surveillance schemes for identifying and reporting specific musculoskeletal conditions. Such systems should provide an analysis by occupation to target those occupations that display disproportionate incidence. New occupations and emerging technologies in which workers may be at risk from exposure to unprecedented biomechanical stresses also need to be identified.

Existing data clearly indicate major deficiencies in surveillance systems for identifying work-related musculoskeletal injuries. Because existing data sources were not designed for surveillance of occupational musculoskeletal injury, they do not separate chronic from acute injuries, and they lack a standard terminology for defining the acute and chronic medical conditions in general.

## Hazard Surveillance

Coupled with needed refinements of health surveillance is a similar need for improved surveillance to define the types and ranges of biomechanical stresses that exist in the workplace. The ultimate value of hazard surveillance lies in prevention because it can provide an early warning of potential cumulative trauma.

Hazard surveillance would make use of ergonomic-type surveys, worksite inventories, or biomechanical profiles of various job conditions to identify the types of biomechanical job demands that pose a risk of musculoskeletal injury. Evidence of workplace-related hazards is particularly important because the hazard may be the only reliable way of classifying a musculoskeletal disorder as work-related.

Without some form of workplace data to identify the presence of biomechanical stress, nonspecific chronic health symptoms, such as joint pain and loss of mobility, may be incorrectly diagnosed as nonoccupational and go unreported. In addition to the obvious value for enhancing the validity of surveillance data, information on sources of biomechanical stress for a given occupation can be used to build models that predict the occurrences of musculoskeletal injuries.

## Essential Elements for Surveillance of Occupational Musculoskeletal Injuries

Guidelines should be set for data collection and diagnostic criteria established for classifying all musculoskeletal disease conditions experienced by workers, whether job-related or not. Definitions should be standardized for characterizing discomforts, injuries, and hazards.

Objective criteria should be determined to differentiate occupationally from nonoccupationally related disorders. The system should also identify the methods of reporting.

Multilevel data bases at national, state, and local levels should be established and supplemented with specific longitudinal epidemiologic evaluations. Multilevel recording is important because the long induction periods for many of the musculoskeletal injuries separate the hazard from the effect. A multilevel health reporting system could track health status across jobs, across geographic relocations, and through retirement.

In general, a positive byproduct of an effective surveillance program would be the renewed awareness within the medical community of the prevention benefits to be derived from a standard reporting system. Such a system would also assist occupational health providers in correctly diagnosing, recording, and treating musculoskeletal injuries.

**Evaluating Cause and Effects**

*Coordinating Scientific Disciplines*

Ultimately any effective strategy geared to prevention must be based on a firm grasp of the factors responsible for the targeted disorder. Analysis and evaluation of relevant surveillance and clinical data are required to uncover the key occupational and nonoccupational risk factors.

For musculoskeletal injuries, the evaluation stage in the prevention process requires the coordination of research in several disciplines encompassing both basic and applied approaches. For example, health professionals should be encouraged to work with engineers and scientists to develop research programs that incorporate each specialty in the design and implementation of an intervention. Once an intervention is designed, hardware manufacturers and users should be brought into the process to fabricate and disseminate new, ergonomically designed equipment, tools, and workplaces.

One of the more pressing needs is for studies that examine the patterns of interacting variables. Methods and measurement techniques also need standardization and better documentation to allow adequate verification and validation so that practitioners can benefit from the development.

*Causes of Low Back Pain*

Evaluating the etiology of low back pain is particularly difficult. Much of what is known about the risk for low back injury is based on epidemiologic data. Two categories of factors that modify the risk of injury have been differentiated from recent surveillance efforts: factors associated with the job and personal factors. Job risk factors include load weight, location, and frequency of materials handling, but often go beyond task design to include psychological factors. Personal factors emphasize age, gender, and strength, but the significance of other personal factors such as fatigue, postural stress, trauma, emotional stresses, degenerative changes, congenital defects, genetic factors, neurologic dysfunction, physical fitness, and body awareness should also be evaluated. The list is far from complete. These personal factors in combination with primary job risk factors make determination of the cause of low back disorders most formidable.

Nevertheless, research on low back pain has progressed beyond the stage of identifying risk factors to include evaluations involving dynamic and three-dimensional modeling and laboratory confirmation. Improved techniques are now needed for estimating tissue forces and pressures noninvasively, both in the laboratory and on the job. Such techniques are necessary to evaluate the stress patterns induced during manual materials handling and to determine to what extent they could be eased through varying different load factors or lifting techniques.

*Causes of Extremity Disorders*

Some progress has also been made in isolating key sources of work-related biomechanical stress for the upper extremities. Biomechanical analyses of hand and arm motions have been useful in specifying points of stress related to symptoms of carpal tunnel syndrome and tenosynovitis. Repetition rate, amount of force required, and postural factors emerge as contributors to these injuries.

In contrast, research on occupational injury of the lower extremities has been limited almost exclusively to the knee. Potential sources of biomechanical trauma to the knee that have been identified include repetitive loading, constant kneeling, squatting, and repetitive contact between the knee and specialized tools.

In general, continued research is needed to combine anatomic, mechanical, physical, and human factors in describing work populations by their abilities to work on a given task without injury. These profiles would provide guidelines for designing new jobs and tools to protect the limbs and joints from excessive biomechanical stress.

**Controlling Occupational Risk Factors**

Three approaches are used for intervention in jobs in which high physical demands pose a risk to the musculoskeletal system:

1. Redesign the job or tools so that demands can be met by a majority of the population.
2. Train workers in techniques to reduce job hazards; i.e., how to lift and how to avoid awkward postures and repetitive motions.
3. Select only those individuals whose work capabilities meet or exceed the high demand.

*Ergonomic Job/Tool Redesign*

*Operating Principle.* The use of engineering techniques has been a basic tenet of occupational safety and health practice for achieving hazard control, in preference to other methods, such as personal protective equipment and safe work practices, which are less reliable and often less effective. The engineering procedure involves modifying task and tools using ergonomic principles to reduce the effects of biomechanical stress.

Unlike the majority of occupational hazards, sources of biomechanical stress seem hidden within the job as specialized patterns of movement or tool use. The relative obscurity of the hazards necessitates that controls be designed into each job identified as having a high risk of musculoskeletal injuries.

This ergonomic approach is based largely on the assumption that work activities that involve less force, repetition, vibration, weight, and forms of static or constrained postures are less likely to cause injuries and disorders. The approach is desirable because it seeks to eliminate potential sources of problems. Ergonomics also seeks to make safe work practices a natural result of the tool and worksite design without depending on specific worker capabilities or work techniques.

The use of ergonomics in job and tool redesign is still in its infancy, largely because an awareness of the science is lacking and not enough cases have been documented showing the effectiveness of these techniques.

However, a few recent demonstration studies have shown reductions in biomechanical stress with ergonomically designed tools, such as the contoured hand grip on pliers for electrical wiring and the specifically shaped knife for poultry processing. Controlled intervention trials are also needed to demonstrate actual reductions in the prevalence of symptoms and injuries.

Initial economic considerations combined with the lack of sufficient substantive evidence for effectiveness have dissuaded many industries from implementing job redesign that might result in reduced injuries and lost work time and ultimately in potential savings.

In contrast, some industries, when faced with staggering workers' compensation costs and rising disability insurance premiums (e.g., those tied to back injuries from manual materials handling), have been motivated to seek out and implement ergonomic solutions. Some have found ergonomic job redesign an effective adjunct to cost-reduction programs. Analyses often demonstrate that ergonomic intervention can reduce musculoskeletal injuries and also contribute to increased productivity. In addition to decreased medical costs, reductions in lost time from injuries and increased worker productivity can be compelling reasons for adopting an ergonomics program. Equally important, ergonomics can form the basis for an improved quality of work life.

*Design Principles.* For some sources of biomechanical stress, implementing an effective ergonomically based recommendation may be difficult. For example, different overlapping sources of biomechanical stress (e.g., vibration, repetition, load, and posture) may contribute in some unknown way to the onset of musculoskeletal disorders. Often no simple, single change can be made, but numerous adjustments may be required to tailor recommendations to the tool, workstation layout, or organization of the task.

*Worker Training/Good Work Practices*

The success of training programs in reducing musculoskeletal injuries has been mixed. Programs in this area range from rudimentary instruction of workers about safety rules and "how to lift" to elaborate programs conducted by Educational Resource Centers (ERCs) for instructing safety and health personnel in the types of neuromuscular disorders associated with repetitive motion.

Studies that evaluate the effects of training in reducing the back injuries associated with lifting have produced conflicting results. One of the most comprehensive and widely reported studies found no evidence that training had any preventive effect. This finding contrasts with the results of a study evaluating training in materials handling for railway workers, which showed annual decreases in the rate of back injury. The training system was developed in cooperation with a rehabilitation clinic. Similarly studies of "work hardening" or physical conditioning programs designed to improve trunk and leg muscle strength for preventing back injuries have largely failed to confirm the long-term efficacy of the programs.

Efforts are under way in some industries to broaden the training beyond the fundamental issues of safe work practices to include the following:

- recognition programs for increasing worker awareness of the hazards
- problem-solving programs designed to provide workers with the information and skills necessary to participate in hazard control activities

Many of these more comprehensive training programs have been supported jointly by management and worker organizations. The federal government has provided some limited assistance in grants from the Occupational Safety and Health Administration (OSHA) as part of their New Directions program.

Such worker training programs are believed to be effective in reducing job-related injuries, but available data needed to evaluate such programs are limited, particularly with respect to preventing disorders.

Training and education of professional safety and health personnel is provided in part by NIOSH, especially through its funding of ERCs. The impact of this training is also difficult to gauge in terms of directly reducing musculoskeletal injuries and illness. The ERC program, however, and its ergonomic and musculoskeletal components have stimulated an increased awareness among occupational health professionals of the basis for occupational musculoskeletal injuries. ERCs have also served the local communities as resources for information and guidance in implementing workplace interventions for controlling such injuries.

Thus, training programs have traditionally focused on teaching employees specific work practices for safety and hygiene. Although most experts support such training, the preventive utility has been difficult to evaluate. The concepts of training have been extended recently to include elements of recognition and problem solving. Moreover, education aimed at awareness is now recognized as necessary at levels of management, including staff specialists, such as tool and workplace designers and engineers.

*Selection/Placement*

Employment screening for musculoskeletal injuries has been used to predict the risk of low back disorders. Selection procedures include anthropometric attributes, such as weight and stature, and the use of back radiographs, muscle strength tests, tests of physical fitness, and tests of lumbar mobility.

The success of any screening program requires accurate information on actual job demands as well as precise measurements of worker capacities related to the key job demands. For example, muscular strength is generally considered an appropriate job-related criterion for work involving manual materials handling. However, measuring the capacities of a worker that most closely reflect the key strength requirements of the job is difficult. Moreover, strength measures are sensitive to many psychological variables, including motivation, expectation, and fatigue tolerance. Studies in which appropriate measurements have been made show a higher incidence of claimed back injuries and back pain in those jobs demanding high exertion in relation to the worker's own maximal isometric strength.

In contrast, the use of anthropometric guidelines for selecting workers has not been justified in studies of manual materials handlers when the outcome was measured in incidence rates for the reduction of low back pain. Although radiologic measurements have been largely discredited as a screening procedure for back problems, they are still widely used. Caution is urged to avoid a potential radiation hazard from overuse. In theory, the assessment of job demands for work output offers a valid alternative for reducing the incidence of musculoskeletal injuries. In practice, this approach is difficult to implement because of the wide variety of demands inherent in the manual jobs of most industries, the range of individual physical capacities, and the lack of criteria for safely matching workers to jobs.

Thus, the ergonomic approach to workplace redesign may be the first choice for controlling musculoskeletal problems, with employee selection and training secondary. Several reasons for this priority exist:

1. Selection and training require that each new employee be evaluated, instructed, and thereafter monitored to determine changes in capacity and compliance with the training procedures. In contrast, health promotion programs combined with ergonomically sound jobs and tools are relatively permanent, and, once implemented, do not normally require modification for each new employee.
2. Employee screening and selection techniques discriminate those considered fit for the job from those who are not. Fitness for a job must therefore be based on actual job demands, which are often difficult to assess. Caution must be exercised that selection procedures are specific to the job, and the general criteria of selecting only the strongest or youngest workers must be avoided.
3. Although training programs are easily implemented and may initially appear less costly than other forms of intervention, they can be more expensive over time because each new employee must be trained and then periodically given a review.

## Increasing Awareness and Stimulating Interventions

### Obstacles to Implementation

The success of any prevention strategy ultimately depends on the potential for implementation. Implementation may be facilitated by increasing both public and professional awareness of the causes and effects of musculoskeletal injuries and by establishing and then conveying an effective rationale for implementing ergonomic or other solutions to these problems. It is difficult, however, to determine the most effective means of getting the information to the public, and how to best use existing communication networks.

### Need for Effective Communication Models

Effective communication is particularly important when the subject involves changing attitudes and behavior. The targets of such communication are either individual workers or organizations (including unions and management or both). Various means of disseminating occupational health information have been proposed, ranging from contracts with local physicians to in-plant clinics or comprehensive health maintenance organizations with occupational health speciality services. Sponsors could include corporations, unions, private practitioners, foundations for medical care, independent medical organizations, and local and state governments.

*Dissemination Needs*

Several dissemination needs can be identified. One pressing need is the provision of ergonomically trained personnel at regional levels such as state health departments, industrial commissions, or similar entities who could offer technical assistance on musculoskeletal disorders to employers and employees. Because only two or three centers of ergonomics currently exist in the United States, the number of regional centers should be increased to provide outreach service and education to local employers. The number and quality of special ergonomic and musculoskeletal courses in regional universities also need to be increased. The number and quality of specialized courses to help workers and employers identify acute and chronic musculoskeletal risk factors should be expanded and the necessary training provided for them to devise their own prevention strategies.

A best method has not yet been determined for increasing the awareness of the musculoskeletal problems within small businesses. A significant number of businesses in the United States employ 25 or fewer workers. These small businesses rarely provide occupational safety and health services, and employers and workers may not recognize the special musculoskeletal health problems created by biomechanical stress.

More in-depth courses on occupational medicine and specific treatment should be included in medical school curricula. The NIOSH Minerva project in the ERCs should continue to encourage the inclusion of occupational safety and health subjects, such as ergonomics, in the curricula of schools of business.

Finally, user-oriented guides for prevention and control of cumulative trauma should be disseminated to workers identified through previous surveillance as being in high-risk jobs.

Other dissemination issues involve defining the role of computer technology and telecommunication networks. How should this technology be used to make information about hazards and potential ergonomic solutions more readily available? What type of public service messages would be effective? What role should occupational and professional organizations play in alerting the public to risk factors for osteoarthritis and to the potential preventive measures involving physical fitness?

## CONCLUSION

The NIOSH prevention strategy (excerpted above) emphasizes task evaluation and tool and workstation design. With implementation of the proposed national strategy for preventing musculoskeletal injuries, the NIOSH program will help to:

- identify high-risk industries and occupations for low back musculoskeletal injuries
- evaluate potential task and personal risk factors that may contribute to musculoskeletal injuries
- develop and validate comprehensive clinical evaluation systems to assess musculoskeletal injuries
- identify and evaluate intervention strategies to reduce the incidence and severity of musculoskeletal injuries
- provide technical assistance and technology transfer concerning musculoskeletal injury control

A critical component of the proposed strategy is the ability of NIOSH to collaborate with appropriate external constituents for implementing the strategy on a national basis. As the NIOSH musculoskeletal strategy evolves, and with the support and assistance of these external constituents, it is anticipated that the frequency and severity of job-related musculoskeletal disorders will be reduced and the quality of working life in industrial America will be enhanced.

---

### NOTES

1. US Public Health Service: *Promoting Health/Prevention Diseases: Objectives for the Nation*, US Department of Health, Education, and Welfare, 1980.

2. US Public Health Service: *Prevention '82*, US Department of Health and Human Services, Office of Disease Prevention and Health Promotion, DHHS publication No. 82-51057, 1982.

3. Millar JD, Myers MD: Occupational safety and health: Progress toward the 1990 objectives for the nation. *Public Health Rep* 1983;98 (No. 4):324–336.
4. *Accident Facts.* National Safety Council, 1985, p 26.
5. Klein BP, Jensen RC, Sanderson LM: Assessment of workers' compensation claims for back strain/sprains. *J Occup Med* 1984;26:443–448.
6. Goldberg H, Kohn T: Diagnosis and management of low back pain. *J Occup Health Safety* 1980;6:14–30.
7. National Institute for Occupational Safety and Health: A proposed national strategy for the prevention of musculoskeletal injuries, in *Proposed National Strategies for the Prevention of Leading Work-Related Diseases and Injuries.* Washington, DC: Association of Schools of Public Health, 1986, pt 1, pp 17–34.

# Part II

# Ergonomics

The studies of ergonomics are diverse and varied. In general, all practitioners of ergonomics desire to decrease work injury and increase work efficiency. The relationships among the worker, the work, and the worksite are important and can be approached in many different ways. For example:

- The engineer or biomechanical expert will view angles, distances, forces, and repetitions in a physics-mathematical model.
- The psychologist will study stress at the worksite from a perceived effort, psychophysical approach.
- The physiologist or medical practitioner will look first at the reactions in the body to exertion, repetitions, in relationship to musculoskeletal and cardiopulmonary functions. This particular point of view is appropriate both for the general work force and for the injured worker.

The sections in Part II, as a continuation of a text written from a physical medicine perspective, emphasize the physiological point of view of ergonomics. It is eclectic enough, however, to include many points of view in the ergonomic spectrum.

The sections in Part II are:

Section 6   Ergonomics and Cumulative Trauma
Section 7   Matching Worker and Worksite—Ergonomic Principles
Section 8   Three Approaches to Specific Ergonomic Problems

# Section 6

# Ergonomics and Cumulative Trauma

*Susan J. Isernhagen*

## INTRODUCTION

### Definitions

Ergonomics has been defined as the scientific study of the relationship between man, work, and the working environment. It incorporates the use of physiological and physical engineering principles to make motion, function, and work safe and efficient.

*The Worker*

Regarding worker safety, one must have an understanding of general principles relating to strength, endurance, aerobic capacity, and tolerance for stressors in the work environment. Individually, people do differ; and, therefore, ergonomic principles, although designed with the "average" person in mind, also must include variations that will accommodate the individual.

*The Work*

Work can be defined as the work motions and activities. It has a strong influence on the safety of the worker. Factors of ergonomic importance are body postures, productivity pressures, musculoskeletal efforts, and physical stress applied to or by the body. Work can be modified by change in position, repetitions, force applied, and task scheduling.

*The Worksite*

The worksite is the physical environment in which the worker does the work. For example, the work station, work tools, and external factors (eg, heat, ventilation, noise) all affect the worker and the work that is being done. The worksite is a physical variable, often one of the most expensive components to fix, but one that must not be overlooked in increasing safety.

## Cumulative Trauma by Any Other Name

Although cumulative trauma disorders have always been part of the medical model, they are increasing in importance today with the improved automation of our industrial sites. Upper extremity cumulative trauma is also on the increase as our society moves from heavy industry into the service industry, which requires more desk work and small assembly or hand machine manipulation.

Cumulative trauma disorders are often categorized by other names, sometimes dependent on regional use or national use. Synonyms for cumulative trauma disorders include repetitive strain injuries, occupational overuse syndromes, and variations using different combinations of those same words.

## Ergonomics—A World Study

In a overview of the physiotherapist in occupational health care at the Australian Physiotherapy Association National Conference, Brisbane, Australia, May 1985, Tuulikki Luopajarvi reported the progress of Scandinavian countries in occupational health. She mentioned that the focus has shifted from treatment to the prevention of health problems in the workplace. Governmental funding is provided either through health acts or through agreements with the labor market organizations. In each of the Scandinavian countries the employer also provides preventive occupational health services. Either municipal or private health facilities are available for consultation.

In Denmark, physical therapists are involved in occupational health and ergonomics involving official labor inspection. This work consists of preventative measures of using education and ergonomics. Regularly scheduled workshops are held for physical therapists in occupational health care.

In Norway, many physiotherapists work full time in occupational health care, including official labor inspection. The main emphasis is on work but curative work often is included. There is a four-week postgraduate course organized by Norway's Institute of Occupational Health, which has a strong component aimed at medical professionals. In addition, courses in ergonomics are available for physical therapists.

Sweden uses physical therapists in a similar manner. Education is provided in the form of four-week courses provided by the government. Physical therapists often work in a team with a safety engineer, and many work within industry.

Finland has a large number of physical therapists working in occupational health care. There are occupational health and ergonomics courses for the physical therapist at the undergraduate level and additional courses in the graduate programs organized by the Institute of Occupational Health.

In summary, Scandinavian countries have found it valuable to use medically based personnel in ergonomics in industry and to provide specific education to this end.

Australia has instituted a national program called "Work Safe," which employs medical professionals in a similar manner to the Scandinavian countries. Also, in Australia, private companies in rehabilitation of the injured workers not only treat the employee-patient but also seek solutions at the worksite. This combination of visits to the worksite and integration of treatment and work has been highly regarded in Australian circles.

In continuing a discussion of ergonomics, Luopajarvi stated that one of the functions of the therapist-ergonomist is to optimize job adaptation at the worksite. The medical professional uses biological and social knowledge to improve technical work means and increase technical solutions. He or she is involved in planning and renovation of worksites, work methods, and work organization. Cooperation is strong with work study engineers, foremen, and work instructors. Communication is very important in implementing the musculoskeletal safety concepts at the worksite.

Although all cumulative trauma disorders have the similarity of being caused by long-term abuse of physiological systems, they differ very much in their remediation. The use of a medical consultant in evaluating the effect of the work and the worksite on the worker is extremely important for proper physiological considerations to be made. In addition, engineers and worksite safety officers should be used in a team approach when industrial suggestions are considered. It will be very cost effective for industry to use a medical consultant in the establishment of protocols to reduce or eliminate cumulative trauma disorders.

## CUMULATIVE TRAUMA AND WORK INJURY

Sudden injury at work is usually clear cut and of an accidental nature. Note only is the cause and effect of a sudden injury easier to document, but the prevention is definable by removing dangerous worksite components or educating against accidents.

A more difficult but very prevalent type of injury is that of cumulative trauma. Many work injuries are reported only at the time that the cumulative trauma becomes intolerable. For example, a worker may report a back injury only when the muscle strain becomes so severe that work has to be curtailed. In fact, however, the injury may have been slowly accumulating and affecting work to some degree for a period before it was reported. Upper extremity cumulative trauma has a similar progression. Aches, pains, and stresses can be noted far in advance of the injury being reported. The current increase in recognition of cumulative trauma now encourages earlier reporting and earlier intervention before a debilitating chronic injury occurs. Understanding the parameters of cumulative trauma aids in application of ergonomic principles.

Results of an injury are visible; the mechanism of an injury is the portion of the cumulative trauma that is unseen. The impact of cumulative trauma can be lessened by looking beyond the effect to the cause.

## Causes of Cumulative Trauma

In studying the reasons for cumulative trauma, each factor must be analyzed individually for ergonomic intervention purposes. However, often stressors interact to cause the disorder; and the effect, then, is greater than any individual contributing factor. Consider the following factors in cumulative trauma:

- repetitive exertions—especially in the low back, neck and shoulder, and upper extremities
- forceful exertions—heavy activities necessary for low back or upper extremity use
- stressful postures—static positioning and back flexion increasing intradiscal pressure, or wrist flexion and deviation creating pressure in the carpal tunnel
- low temperatures—causing physiological stiffness in the soft tissues including tendons and ligaments. Cold also decreases circulation and sensation
- vibrations—decreasing circulation or causing microtrauma to soft tissue
- poorly designed workstations—stressing employees by putting them in potentially injurious positions and also by creating a mechanical disadvantage for the worker, creating higher degree of intensity to get the work done

- effect of gravity—antigravity activities that are very strenuous, requiring muscular effort to overcome gravitational force.

Symptoms of overuse injury tend to appear weeks, months, or sometimes even years after continuous performance of the activity. A work history and evaluation of work design, posture, and biomechanics are essential. The occupational history should include:

1. Previous occupations and repetitive duties
2. Previous symptoms
3. Work duration before the onset of symptoms
4. Anatomical pattern of symptoms
5. Type and duration of symptoms after initially noted
6. Treatment effects
7. Work stress factors identified above
8. Outside activities that may aggravate symptoms

**Types of Cumulative Trauma**

A representative list of specific types of cumulative trauma are noted here. It is not a complete list but serves to exemplify the types of injuries for which ergonomic intervention is necessary.

*Wrist/Hand Conditions*

Carpal tunnel syndrome is a condition where structures in the carpal tunnel of the wrist are compressed mechanically or by swelling. It is often caused by repetitive work with the hands that would require repeated wrist flexion, forceful ulnar or radial deviation, repeated forces on the base of the palm and the wrist, and forceful pinching.

deCuervain's syndrome is characterized by tenderness and pain over the upper portion of the wrist on the thumb side. It is caused by inflammation of the muscles of the thumb that are aggravated by high repetitions during manipulation, blunt trauma, or repeated ulnar deviation of the wrist with forceful motions of the thumb.

Ganglion cysts are hard, small areas of fluid in the tendons of the wrist or fingers. They may be caused by sudden or strong use of a tendon joint. They often are associated with repeated manipulations with extended or deviated wrists and forceful grip.

Epicondylitis (eg, "tennis elbow") and related injuries are caused by continued stressful use of wrist-hand extensor or flexor muscles. The continued use of wrist extensor or flexor muscles often is combined with forearm pronation or supination.

### Neck/Shoulder Conditions

Forward head position may exist in employees who have faulty posture. It also can be accentuated by work that causes forward bending of the head and back during desk or assembly line activities. One outcome is habitual stressful postures that lead to tightness and weakness of postural muscles.

Rounded shoulder posture and/or thoracic outlet syndrome is aggravated by work that brings the arms into a forward position, with head and neck in a rounded position. It is characterized by pain and numbness in the arms. Poor lighting may accentuate the problem.

Tension/stress problems in the neck and upper shoulders are a byproduct of prolonged work without rest. Tension due to stress at work, which includes pacing, may cause continued contraction of the muscles even when work is not being done. This increases the likelihood of decreased circulation and increased nerve problems. Fibrositis and myositis may be chronic outcomes of this condition.

### Low Back Dysfunction

Sprain and strain of the low back includes low grade pain accompanied by static positioning, usually in the flexed posture, or high repetitions of movement of the low back. It is characterized by discomfort and, in the more severe forms, muscle spasm. It is aggravated by activity, and symptoms decrease with rest.

A bulging disc syndrome can be caused or accentuated by extremes of back flexion or extension. Flexion postures causing posterior disc bulges are the most common causes. Examples are prolonged sitting and work activity with a rounded back.

### Lower Extremity Dysfunction and Joint Problems

Degenerative changes can be a combination of aging changes and mechanical stress on lower extremity joints, such as the hips, knees, and ankles. Cumulative trauma will be noted when there is extreme force or repetitive movement to the joints. This can happen in compression-type activities, such as jumping, running, loading with joint twisting, and climbing.

### Decreasing the Potential for Cumulative Trauma

When cumulative trauma has been identified in an individual worker or work group, the following options can be beneficial in decreasing trauma potential:

- Review all reports of cumulative trauma injury to look for patterns. Worksites that exhibit more than one case of cumulative trauma should be evaluated for potential work and worksite modifications.
- Stress stability rather than mobility of the body when possible. Then ensure that stability is alternated with exercise to prevent static posture problems.
- Use work rotation. Rotate tasks frequently within any particular job, and implement rotation of workers.
- Keep objects close to the center of gravity of the body when lifting or moving them.
- Assure that handgrips are the optimal size for each worker. Adjustable handgrips are helpful.
- Use the workers as a problem-solving group to assess the ergonomic situation and suggest solutions.
- Mandate regular breaks including rest or exercise at lunch time.
- Allow each worker options for self-pacing.
- Allow each worker options for changes of position.
- Use gravity to assist rather than resist motion. Avoid antigravity static positions, such as shoulder abduction or hip flexion, while standing. Provide support when antigravity static positioning is necessary, including elbow and wrist supports for upper extremity activity.
- Keep reaching to a minimum.
- Avoid sharp edges on work surfaces.
- Allow ample leg room for both sitting and standing activities. Provide foot support for alternating foot positions, especially when standing.
- Allow a balance to take place in standing and sitting postures.
- Evaluate muscle loading for both strength and endurance factors.
- Use padded surfaces, including padded chairs and floor mats, for standing.
- Assure the chair fits properly—with low back support to maintain the normal lordotic curve and height adjustment so that feet rest comfortably on floor or footrest.

- Allow 90° flexion at the elbow for sitting and hand-coordination activities.
- Diminish or avoid vibration when possible. Use vibration absorbers for both hand and sitting activities.
- Keep temperature in the work area as moderate as possible.
- Assure relaxation of muscle groups occurs after each contraction. This avoids constant contraction, which decreases circulation.
- Avoid sudden impact, such as using the hand as a hammer, jerking, or sudden start-stop movements.
- Use tools that allow joints to stay in a neutral position. This would include bent-handled hand tools and placement of work parts in the proper position.
- Use activities that involve low repetition, low weight as a warm-up and low repetition, low weight as a cool-down after periods of heavy activity. (See Pause Gymnastics—Exercise Breaks in Section 4-2, *supra*, for recommendations on both flexibility and strength.)
- Evaluate effect of gloves on gripping ability.
- Avoid preshaped tools, such as handgrips with finger grooves. These may be less appropriate for the majority of people.

All of these precautions will be very helpful when combined with employee education. This education should be designed to allow workers to use their body in the most effective manner. They also should be encouraged to report symptoms as early as possible.

### Individual Susceptibility to Cumulative Trauma

Table 6-1 provides a checklist that can be used by medically oriented ergonomists in several types of ergonomic/cumulative trauma evaluations. By going through this physiological checklist, an expert could discover precursors to injury (ie, the person's predisposition or susceptibility to injury), or the cause of injury after it is reported. The information can be used for individual cases or designed for groups of workers.

**Table 6-1** Checklist of Factors To Consider in Evaluating Physiological Ability for Work

### Muscle

*General considerations:*
- strength capacity
- endurance capacity
- synergistic movement
- eccentric or concentric contraction
- speed of contraction
- length of muscle during maximal contraction
- blood flow restoration during rest

*Individual considerations:*
- relationship of strength fibers to endurance fibers (jogger versus sprinter)
- strength
- endurance
- strength of agonist to antagonist
- coordination
- ability for stabilizing
- posture

### Tendon

*General considerations:*
- tendon sheath and lubrication
- angle of bend during motions
- presence or absence of inflammation
- entrapment
- amount of tearing stress

*Individual considerations:*
- proneness to rupture (strength, age, and use)
- type of bony prominences
- small-large joints and joint openings
- metabolic and hormonal conditions

### Joints

*General considerations:*
- synovial fluid
- cartilage smoothness
- motion at mid-ranges
- back—presence or absence of:
  weighted rotation
  prolonged positioning
  proper body mechanics

**Table 6-1** *continued*

*Individual considerations:*
- joint mobility
- arthritic changes
- size of joint
- length of bones for lever arm

**Circulation-Cardiopulmonary**

*General considerations:*
- blood flow during rest between contractions
- blood pressure
- pulse

*Individual considerations:*
- tidal response volume (conditioning, age, and disease)
- presence of diabetes mellitus
- aerobic capacity

## Anthropometric Information

In the context of cumulative trauma, anthropometric data are a collection of measurements of large numbers of humans to establish average size ranges. Once average height, reach distance, hip width, hand size, and so forth, are known, the design of work stations and work becomes more scientific. Establishing of norms of activity allows the greatest percentage of the working force to be within safe parameters in design of work and work spaces. Although anthropometrics does not allow for full human variability, it does allows us to formulate a basic plan for design of work stations and work and then make individual adjustments.

Much of the early anthropometric data were gained from studies of the military. Therefore, early designs were often accommodated to younger men. Because the work force changed, including an increase in the number of women and older workers, anthropometric data needed to expand to include all categories of people. The following are guidelines to aid in analysis of such data for your purposes. Match your needs—ie, the population group you are studying—to the group of subjects from which the data were taken.

- What is the proportion of men and women in the study? Are values given for men and women as separate groups or are they averaged

together? Generally speaking, they should be given as separate groups if both genders are involved in the work and worksites. Neither group should be disadvantaged, so parameters for both men and women are needed.
- Were all age groups represented? If so, again, one needs to know whether all values are averaged or whether age category information is obtainable from the data. Again, it is better to be able to establish age-related data by group.
- What is the extent of the ranges? For example, if you are measuring for work stations, do the ranges include the middle 75% of the population, or do the ranges include the middle 10% of the population? This will tell you how many people will fall out of the intermediate categories given.
- Are all levels of subjects included? If the general public is measured, the data will be amenable to the general population. Many studies, however, use a college population, a military population, or a working population. The values then can be generalized for that population but not for humans as a whole.

An example of the use of anthropometric data is illustrated by common recommendations for proper seating in workplace design:

- Chair seat height: adjustable, between 16 and 22 in
- Sit-stand work station: adjustable, between 27 and 44 in
- Adjustable footrest: adjustable, between 11 and 14 in
- Seat size: 16 × 16 in to 20 × 20 in
- Seat slope: ±6° about horizontal axis
- Back rest height, lower edge: adjustable, between 3 and 6 in from the seat

Work with the use of anthropometric data is particularly valuable to engineers and industrial designers who plan work stations. The data used in design should be correlated with the work force currently at the worksite as a self-selection process may have been present at the time of hiring. For example, 95% of the workers may be men. Other self-selection variables within a work situation might be height, weight, age, or strength categories. The more individualized the data on the work force are, the more the proper anthropometric data can be selected.

# Section 7

# Matching Worker and Worksite—Ergonomic Principles

*Suzanne H. Rodgers*

The major goal of ergonomics is to design jobs, equipment, environments, and products within the capabilities of most people. If this is done, it is not difficult to return people to work after an illness or injury. However, this goal is far from being realized, and there are many jobs that are too difficult physically for many eligible workers to perform for a full shift. In most of these jobs either worker self-selection or "natural selection" is used to find the people who can do the work. Those who can perform the tasks safely often tend to be young, male, and physically fit.

Although this trial and error method can be used to find people to perform difficult jobs, often a side effect is high job turnover and strain and sprain injuries are common. The ergonomist can approach such jobs in two ways. One approach is to define the job requirements and the capabilities needed to perform them so that a person can be selected for the job based on a measurable strength or endurance capacity. The second approach is to define the job requirements, identify the limiting activities, and determine how to lessen those demands so that more people can do the job. Selection testing does not solve the job problem and requires continuous screening of applicants for the job. Job redesign eliminates the problem and is probably the less expensive alternative over time, although it may require a significant up-front expenditure. For this reason, the second approach will be discussed in detail in this section.

## CHARACTERIZING THE JOB DEMANDS

To determine the suitability of a job for a person who is coming back to work after an illness or musculoskeletal injury, there are three primary requirements to characterize:

1. The intensity of effort required.

2. The amount of time the effort has to be sustained continuously (duration).
3. The pattern of effort exertion over the total working period, or shift.

Other factors, such as environmental conditions, social interactions, and psychological stresses, are also important but will not be discussed here. A model of upper extremity disorders, especially those of the hand and wrist, will be used to illustrate an ergonomic approach to matching the job and worker and to suggest ways of improving jobs that are outside the capabilities of many people.

**Effort Intensity**

The strength required to exert force or lift something will depend on the muscles used and their posture during the task. For work with hand tools, the amount of force exerted on the tool should be identified and the type of grip the person can use for the task should be characterized. The force on the tool can be measured directly using a torque meter.[1] If this is not available, reasonably good estimates can be made by asking the worker to recreate the force while pressing on a platform scale (eg, bathroom scale) or pulling on a spring scale (eg, fish scale) using the same posture as is used in the task.

The type of grip used can be defined as either a power grip or a pinch grip. Power grips involve encirclement of the object by the fingers and thumb, similar to the grip used to hold a hammer. Pinch grips involve holding objects between the thumb and finger tips or between the thumb and the side of the index finger (lateral pinch). The primary difference between power and pinch grips is that pinch grip strength is only about 20% to 25% of power grip strength.[2] Consequently, a required force of 15 lb (or 150 N) could be perfectly acceptable for most people if they are using a power grip (requiring about 25% of their strength or less) but would be a maximum exertion if they have to use a pinch grip.

Other factors limit the amount of power grip strength available, especially grip span and wrist postures.[3] Too large or too small a grip span make it difficult to curl the fingers and bring them together in a power grip. The grip muscles in the forearm have to exert high forces to get the required force at the hand in these biomechanically inefficient postures. Wrist posture also affects the biomechanics of grip exertion by making the tendons from the muscles in the forearm pull around corners instead of directly on the fingers and thumb. This means that larger forces have to be developed in the forearm muscles to get the same strength output that can

Matching Worker and Worksite   67

**Grip Strength and Wrist Angles**

| Wrist Position | % Neut Grip |
|---|---|
| Rad 30 | 72 |
| Rad 15 | 81 |
| Uln 45 | 62 |
| Uln 30 | 74 |
| Ext 60 | 63 |
| Ext 45 | 76 |
| Flex 75 | 38 |
| Flex 60 | 48 |
| Neutral | 100 |

**Grip Span and Strength** *

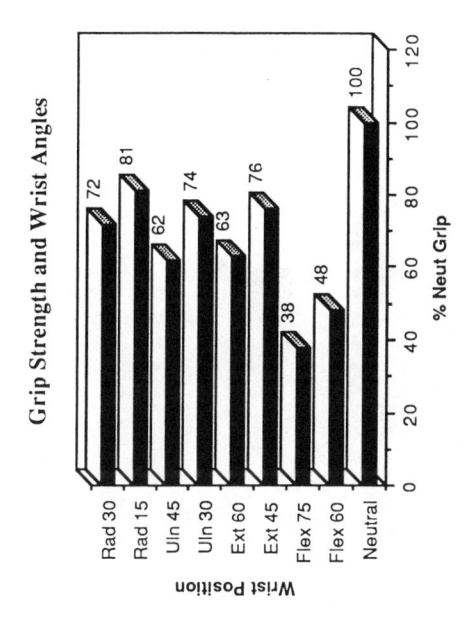

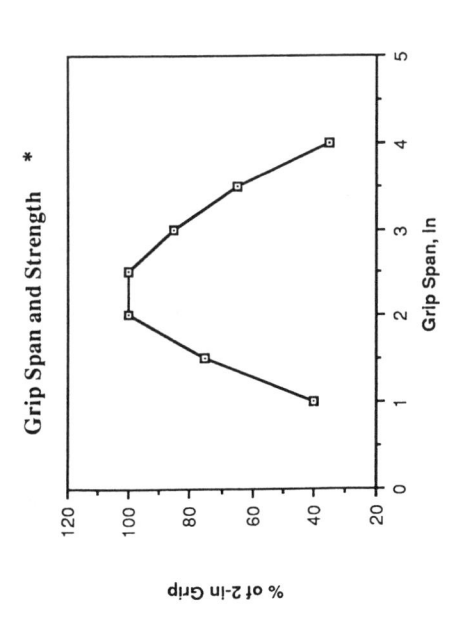

(% of 2-In Grip vs. Grip Span, In)

Flex = flexion; ext = extension; uln = ulnar; rad = radial; neut = neutral.
*Grip over two parallel surfaces.

**Figure 7–1** The effects of grip span and wrist angles on power grip strength.

## Gloves and Grip Strength

[Bar chart showing Grip % of BH: Bare-Handed = 100, Rubber = 81, Cotton = 74, Heat-Resist = 62]

BH = bare-handed.

**Figure 7-2** The effects of gloves on power grip strength.

---

be achieved when the wrist is in the neutral position (no flexion, extension, or radial or ulnar deviation). Figure 7-1 (based on Eastman Kodak Company[3] and a conversation with M.J. Wang and R. Rabadi, October 1987) shows how grip span and wrist angles influence power grip strength.

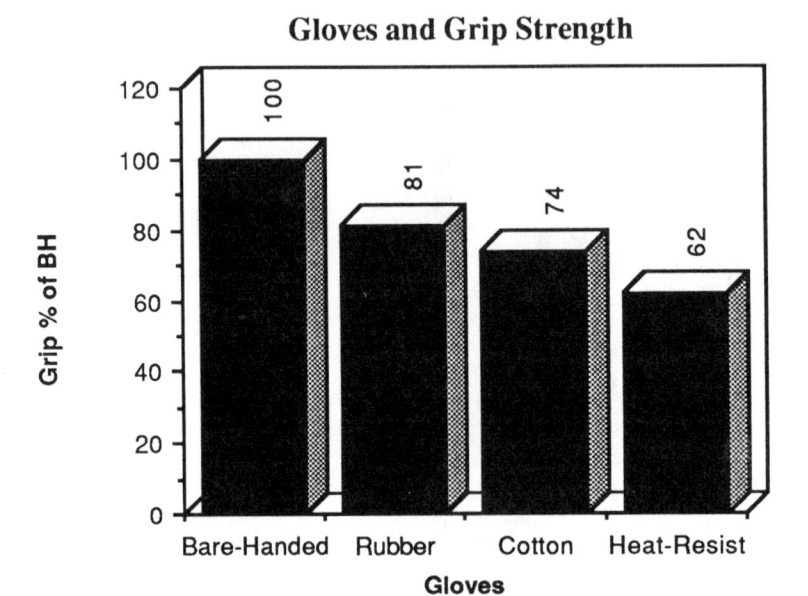

Another factor that influences the amount of power grip strength available on a job is whether or not gloves are worn. Thin gloves without seams between the fingers interfere the least with grip force development. Thick gloves with seams can reduce grip strength by up to 40%. Figure 7-2 summarizes the effects of wearing gloves on power grip strength.[4]

Having estimated the force exertion requirements of the job and determined the type of grip, wrist postures, hand span, and characteristics of any gloves being worn, the job requirements can be expressed in terms of a percentage of available strength for most people. This is referred to as the percentage of maximum voluntary contraction (% MVC) strength, for the muscles most involved. For example, about 75% of the potential work force can be expected to have a maximum power grip strength of at least 60 lb (or 27 kg) if the grip span is from 2 to 2.5 in (5 to 6.2 cm), the wrist is in the neutral position, and no gloves are worn.[5] So, a force of 20 lb exerted in a task would use 33% of the capacity of a person with a 60-lb grip strength. If the task is designed so the worker has to flex the wrist 45°, the available strength would be 75% of the original value, or 45 lb (about 20.5 kg); the same 20-lb force would now be 44% of capacity. If the tool being used has a 1-in (2.5 cm) grip span instead of a 2 to 2.5 in (5 to 6.2 cm) span, power grip strength would be 40% of the initial value or 24 lb (11 kg). With a neutral wrist and no gloves, the 20-lb force is now 83% of capacity. When awkward wrist postures, gloves, and suboptimal grip spans are all present, only those people with high grip % MVCs will be able to perform the tasks on a repetitive basis. (See Figure 7-3.)

## Duration of Continuous Effort

A most important concept for ergonomic design is the relationship between effort intensity and duration for continuous work; that is, not only how intense an effort is needed to perform a task, but how long that effort must be sustained before there is a break to allow the muscles to recover. As the effort intensity increases, blood flow into the working muscle is compromised. At about 40% MVC blood flow reduction is measurable, and at 70% MVC there is a very significant reduction.[6] Because the blood carries oxygen which is vital for supplying energy for muscle contraction, the higher the % MVC used, the shorter the time it can be sustained continuously (Table 7-1).[7,8]

**Table 7-1** Percentage of Maximum Voluntary Contraction Strength versus Continuous Holding Time

| % Maximum Voluntary Contraction Strength | Continuous Holding Time (s or min) |
| --- | --- |
| 100 | 6 s |
| 85 | 12 s |
| 70 | 20 s |
| 50 | 60 s |
| 40 | 2.5 min |
| 25 | 3.5 min |
| 15 | >4 min |

**Figure 7–3** An example of work affected by the use of gloves. Analysis of grip strength, wrist angle, and grip span is useful in this example.

Without enough oxygen to supply the energy for muscle contraction, anaerobic metabolism is used and lactic acid is formed. This byproduct of inadequate oxygenation has to be sent to the liver to be reconverted into glucose or glycogen, so it takes longer to recover from a period of $O_2$ debt in the muscles than it does to develop the debt in the first place. The availability of recovery time between intensive exertions determines the acceptability of some highly repetitive tasks. As the frequency of repetition increases, there is less recovery time between exertions. Even if the effort is quite short (3 to 5 s), there may not be enough time to recover if the intensity of the effort is high. Figure 7-4 shows the needed recovery time for efforts of several different intensities as a function of holding time.[9] By summing the holding and recovery times for a given effort intensity and dividing it into 60 s, the best frequency for performing the task can be determined.

The duration information can be used to identify whether the work will be suitable for most people to perform in a repetitive manner. If the effort has to be sustained for long periods (one or two minutes), the duration information will determine whether most people can perform the task even on a lowly repetitive basis (see Table 7-1). Ways to improve the tasks to make them suitable for more people are discussed later.

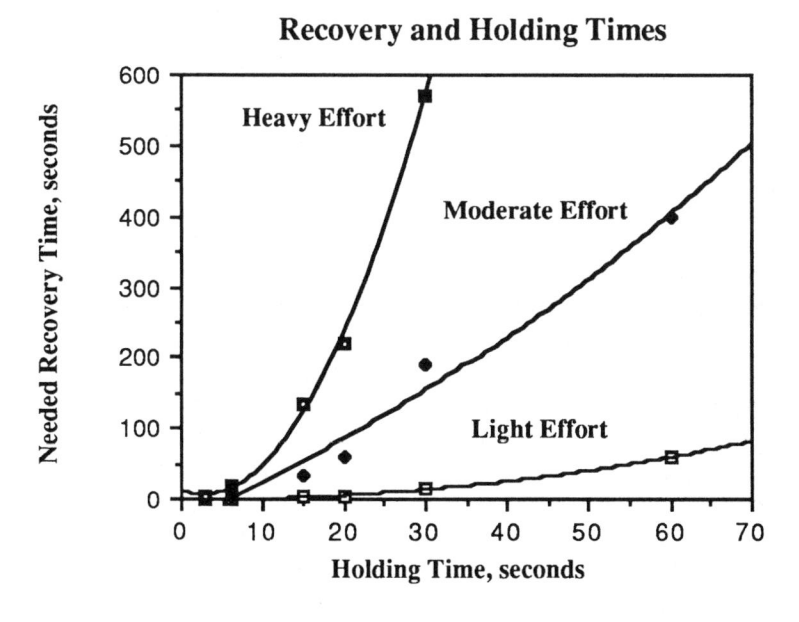

**Figure 7-4** Recovery time needed at various intensities as a function of holding time.

## Work Pattern

If the effort intensity and duration requirements of the job are appropriate, still to be considered are the overall job demands for the shift. A person may be able to sustain a task for 30 to 60 minutes but may reach limits when trying to sustain it for 120 minutes without a work break. It is important to observe the degree of control that the worker has over his or her work pattern. The more control, the easier it is to get the worker back to the job, because he or she can arrange the work to fit his or her needs for recovery time.

Intermittent work, with short (one- to two-minute) breaks alternated with intense efforts, is more efficient as a work pattern than sustained intense work, with a 15- to 20-minute rest break afterward.[10] If a machine or other outside influence is determining the work pattern and requiring the worker to match its pace, the worker with a lower capacity will be at a disadvantage.

Work patterns are best categorized by describing the distribution of tasks in time and over the shift, then timing within the major tasks to determine if periods of lighter effort are interspersed with heavier efforts. In addition, it is useful to observe specific muscle groups and to identify their work and recovery cycles. Videotaping the jobs permits these analyses to be done more carefully. Several people should be taped on the job so individual differences in technique can be appreciated.

## CHARACTERIZING THE PERSON'S CAPABILITIES FOR THE WORK

In the discussion of job demand characterization, it was mentioned that physical job demands can be expressed as percentage of capacity for given muscle groups. There are data available on the strengths of specific muscle groups in men and women and on the strength available for pulling up or pushing and pulling horizontally when several muscle groups are used. Some of the relevant strength measurements are given in Table 7-2,[3, 5, 11, 12] which includes average values and values representing weaker people (5th and 25th percentiles) and stronger people (75th and 95th percentiles).

To determine if a person has adequate strength to do a task, the active muscle groups should be identified and the patient's strength should be tested in the posture used on the job. In the job analysis discussed earlier, the job strength requirements were measured; from this point, then, job requirements can be compared to the strength available, expressing the job

**Table 7-2** Muscle Strength Measurements

| Measurement | Sex | 5th | 25th | 50th | 75th | 95th |
|---|---|---|---|---|---|---|
| Power grip | F | 24 | 51 | 60 | 69 | 96 |
|  | M | 39 | 85 | 101 | 117 | 163 |
| Elbow flexion | F | 5 | 28 | 36 | 44 | 67 |
|  | M | 10 | 49 | 62 | 75 | 114 |
| Vertical pull up | F | 27 | 50 | 58 | 66 | 89 |
| at knee height | M | 35 | 62 | 83 | 103 | 138 |
| Horizontal push/ | F | 43 | 51 | 59 | 68 | 76 |
| pull at chest height | M | 54 | 62 | 70 | 79 | 87 |
| Acceptable weight | F | 28 | 33 | 37 | 42 | 47 |
| low lift, pounds | M | 37 | 45 | 54 | 63 | 70 |
| Acceptable weight | F | 25 | 29 | 34 | 38 | 42 |
| center lift, pounds | M | 34 | 43 | 53 | 62 | 71 |
| Acceptable weight | F | 24 | 26 | 29 | 32 | 35 |
| high lift, pounds | M | 29 | 39 | 49 | 59 | 68 |

Column header: Muscle Strength (lb) — Percentile

demands as a percentage of the patient's capacity. For example, if an upward pull of 35 lb (16 kg) is needed and the patient has a pull capacity of 70 lb (32 kg) at that location, the job requires 50% of his or her capacity.

If the task is repetitive, the demands have to be related to the work/recovery cycle guidelines presented in Table 7-1 and Figure 7-4. To test the worker's endurance capability, it is possible to have the strength exertions repeated while he or she is in the job for a period of three to five minutes to observe his or her tolerance of repetitive work. If several different types of work are done, it may be necessary to test the strength of several muscle groups to assess the worker's suitability for the work.

For most of the strength measures, a strain gauge or dynamometer is adequate to assess the strengths of interest.[3] Use of a tray, with calibrated weights added, may be appropriate to evaluate a person's capacity for repetitive lifting tasks. The primary concern is to make the test similar to the job requirements by assuring that the same muscles and similar postures are used.

If the job requires frequent lifting throughout the shift, total workload limits may be of concern, so determination of aerobic capacity may be needed. Then the estimated or measured energy requirements and the job

**Table 7-3** Percentage of Aerobic Work Capacity versus Continuous Duration of Work

| % Aerobic Work Capacity | Continuous Duration of Work (min or h) |
|---|---|
| 100 | 6 min |
| 85 | 12 min |
| 70 | 40 min |
| 50 | 60 min |
| 40 | 4 h |
| 33 | 8 h |
| 28 | 12 h |

pattern information can be used to assess what percentage of aerobic capacity is required to do the job. Table 7-3 shows the aerobic workload time limits for different percentages of capacity.[3]

## EXAMPLES OF TECHNIQUES TO REDUCE THE DEMANDS OF HEAVY JOBS

If it is determined that the worker does not have adequate capacity for the job, simply setting a restriction against doing that work should not be done without looking for ways to modify the job effort requirements. Ergonomists will look for ways to reduce the strength requirements or to shift them to stronger muscle groups, to reduce the overall effort levels, and to give the worker control, in so far as it is possible, of the work pattern. Ergonomics interventions in the workplace are discussed below using low back pain and carpal tunnel syndrome as examples of common musculoskeletal problems for which a person may need work modifications.

### Low Back Pain

People with low back pain may have difficulty with jobs that frequently require awkward postures and with repetitive or awkward manual handling tasks.[13] The postural characteristics of jobs that may produce problems include:

- working over shoulder height
- bending over
- crouching down

- constant standing
- constant sitting, especially if the seating is poor
- twisting of the trunk
- extended forward reaches

These can be modified by:

1. Adjusting working heights to reduce bending over or crouching down.
2. Using platforms (when possible) to bring the person up so the work is not done above shoulder height.
3. Orienting the workplace so trunk twisting is less likely to occur (eg, 90° instead of 180° angles between the workbench and supplies).
4. Including activities in the job that allow the worker to get up and walk a bit in a sitting job, or to sit down for short periods in a standing job.

All of these approaches can provide relief for the person with low back pain. Handling aids, such as reach extenders, back supports for chairs, and foot supports, also are useful aids.

Manual handling tasks may make it difficult for a person with low back pain to return to work, but the ergonomist can reduce these problems by using the following techniques[13]:

1. Eliminating barriers that prevent the person from getting close to the load, such as inadequate foot and leg clearances around items to be lifted.
2. Raising items to be lifted to 20 in or more above the floor by using platforms or lift tables, so less energy is needed to lift repetitively and there is less strain on the back.
3. Looking for ways to handle some items in bulk to reduce the amount of manual handling required.
4. Providing less heavy tasks as part of the job to assure adequate recovery time between lifting tasks.

Exhibit 7-1 is an example of a theoretical return-to-work situation that illustrates how ergonomics techniques can be incorporated into the medical evaluation of the worker's ability to return to the job after an attack of low back pain.

Exhibit 7-1 Ergonomic Interventions for Return to Work after an Episode of Low Back Pain

> **The Job:** Packing boxes of product, labeling and sealing them, and then placing them on pallets for shipping. Boxes are 12 × 12 × 12 in (30 × 30 × 30 cm) and weigh 30 lb (14 kg) each. The packing line is 30 in (75 cm) high, the labeling and sealing line is 25 in (62 cm) high, and the pallets are 5.5 in (11 cm) high as they sit on the floor. The lifting rate is nine boxes per minute and is sustained for two to three minutes at a time. In a full shift there are 540 boxes handled per shift on the job. The pallet is loaded five tiers high with 12 boxes on each tier.
>
> **The Worker's Status:** Fifth attack of low back pain in 8 years; some chronic pain, radicular—right leg; had pain when lifting 30 lb (14 kg) at 15 in (38 cm) above the floor and 10 (25 cm) in front of the ankles in a laboratory simulation of the job. No pain on handling up to 55 lb (25 kg) between 30 and 40 in (76 and 102 cm) above the floor and close to the body. Additional pain generated with lifts of 25 lb (11.5 kg) at 50 in (127 cm) above the floor or higher.
>
> **Initial Evaluation:** It appears that the worker cannot go back to this job because it involves frequent handling below 15 in (38 cm), some sustained bending for the labeling and sealing activity, and some lifting above 50 in (127 cm) to stack the fourth and fifth tiers of the pallet.
>
> **Ergonomics Interventions:** Among the approaches that could be considered to reduce the load on the worker's back and to make the job better for most people would be:
> 1. Raise the labeling and sealing conveyor so the labels are applied between 35 and 45 in (88 and 114 cm) above the floor.
> 2. Place the pallets on levelators, so the low lifts are avoided; and provide a platform for the workers to use when loading the top tiers.
> 3. Consider eliminating the fifth tier of the pallet pattern.
> 4. Evaluate the feasibility of sending the boxes to a central area where an automatic palletizer may be cost justified so that hand palletizing is eliminated altogether (long-term solution).

## Carpal Tunnel Syndrome and Tendinitis

People with cumulative trauma disorders of their hand, wrist, and forearm may find it difficult to return to a job on which the symptoms first appeared. Several of the risk factors for these problems were discussed under job demand characterization. These include:

- high grip forces
- awkward wrist angles
- too large or too small a pinch grip span
- use of gloves

In addition, cumulative trauma symptoms are more prevalent in jobs where the following conditions apply[14,15]:

- time pressure
- work with arms elevated, tension in shoulders
- constant holding tasks
- forceful rotations
- cold temperatures
- vibrating tools used
- high frequency tasks (seven or more strong exertions per minute)

To reduce the risk of aggravating the symptoms in people with a susceptibility to carpal tunnel syndrome or to tendinitis, the ergonomist looks for ways to improve wrist and hand postures through workplace and tool design; to relieve neck and shoulder tension by adjusting workplace heights; to improve the grip type and span; to choose appropriate gloves; to reduce the forces required; to provide short recovery periods in constant holding tasks or provide aids to hold or support parts or tools; to control cold and vibration exposures; and to design the job to provide periods of less forceful exertions so the arm and hands can recover and the worker can regulate his or her work pattern as needed to minimize the symptoms.

An example of how this information can be used to help a person return to work after an episode of carpal tunnel syndrome or tendinitis is shown in Exhibit 7-2.

Each of these changes would benefit not only the worker with lowback pain or carpal tunnel syndrome or tendinitis, but all workers. Some will take longer to implement than others, but often they can be phased in slowly as workplace and job changes evolve.

## SUMMARY

The current return-to work process for injured workers who have chronic musculoskeletal problems is directed toward looking for a job that they can do, given their reduced functional capacities. Trying to find the right job creates new problems, such as having to "bump" people with less seniority from their jobs to accommodate the injured worker, having to train the injured worker to do the new job and train the "bumped" worker on a new job as well, having to deal with other psychosocial aspects of creating jobs for injured workers, and so forth. To avoid these disruptions and to improve problem jobs for all workers, it is important to address the risk

**Exhibit 7-2** Ergonomic Interventions for Return to Work after an Episode of Carpal Tunnel Syndrome

> **The Job:** Using a pistol grip power tool to attach nuts to bolts on a product on an assembly line. The nut driver's handle is 3 in (8 cm) wide and the trigger is another 0.5 in (1.2 cm) from the front of the handle (span = 3.5 in (9 cm)). In this job there are eight nuts to tighten on each part and there are 480 parts done per shift. The part is bulky and must be held in place with one hand while the other hand operates the tool. The nut driver is air-driven with the hose connected to an air outlet 30 in (76 cm) from the front of the workbench. When each nut is tight, there is a "kick," which must be resisted to control the tool. The assembly is done at a seated workplace with a 30-in (76-cm) high work surface and a stool that adjusts from 20 to 30 in (51 to 76 cm) above the floor. Incentive pay creates some time pressure.
>
> **The Worker's Status:** Bilateral carpal tunnel syndrome diagnosed using a nerve conduction velocity test. Night pain and aggravation with flexed wrist posture and with pinching tasks or high repetition rate. Grip strength with 2.5-in (6.2-cm) span is 55 lb (25 kg). Symptoms also aggravated when shoulders are elevated.
>
> **Initial Evaluation:** The risk factors are high forces (the "kick"), high repetition (about 8/min throughout the shift), shoulder tension (it is necessary to work with the elbows elevated because of the work surface height and lack of leg clearance), vibration from the tool, too wide a grip span for triggering the tool and for holding the part down with the other hand, and some pacing associated with the pay plan. It does not appear to be a good job for this worker.
>
> **Ergonomics Interventions:** The following approaches would reduce the risk factors for repetitive motions problems on this job:
>
> 1. Reduce continuous holding of the tool and the part by supporting the tool from overhead and providing a fixture to hold the part in place.
> 2. Consider using an in-line (straight) tool to press vertically down on the nuts from above so that awkward wrist angles can be avoided and less "kick" is translated to the hand at the end of each nut tightening cycle.
> 3. Consider using a nut driver with multiple heads to speed up the task and provide more recovery time between parts.
> 4. Use a vibration damping material on the tool handle.
> 5. Improve the workplace seating so the shoulder tension is reduced.

factors that make it difficult for people with chronic musculoskeletal problems to return to their job. The probable results of ergonomic redesign will be to reduce the amount of time lost and restricted from the job, to keep trained people working productively, and to reduce the number of people who develop cumulative trauma or low back symptoms that interfere with their ability to do their job.

## REFERENCES

1. Eastman Kodak Company, Human Factors Section: *Ergonomic Design for People at Work*. New York, Van Nostrand Reinhold, vol 1, 1983.

2. Jacobsen C, Sperling L: Classification of hand grip. A preliminary study. *J Occup Med* 1976; 18:395-398.

3. Eastman Kodak Company, Human Factors Section: *Ergonomic Design for People at Work*. New York, Van Nostrand Reinhold, vol 2, 1986.

4. Wang MJ, Bishu RR, Rodgers SH: Grip strength changes when wearing three types of gloves. Read at HF Interface '87, Western New York Human Factors Society, Rochester, NY, May 1987.

5. Kamon E, Goldfuss A: In-plant evaluation of the muscle strength of workers. *Am Ind Hyg Assoc J*, 1978; 39:801-807.

6. Lind AR, McNicol GW: Circulatory responses to sustained hand-grip contractions performed during other exercise, both rhythmic and static. *J Physiol (Lond)* 1967; 192:595-607.

7. Rohmert W: Problems in determining rest allowances. Part 1: Use of modern methods to evaluate stress and strain in static muscular work. *Appl Ergonomics* 1973; 4:91-95.

8. Scherrer J, Monod H: Le travail musculaire local et la fatigue chez l'homme. *J Physiol (Paris)* 1960; 52:419-501.

9. Rodgers SH: Recovery time needs for repetitive work. *Semin Occup Med* 1987; 2:19-24.

10. Simonson E; Recovery and fatigue. Significance of recovery processes for work performance, in Simonson E (ed): *Physiology of Work Capacity and Fatigue*. Springfield, Ill, Charles C Thomas Publisher, chap 18, 1971.

11. Day DE: *Comparison of Dynamic Lifting Forces at Differing Heights and Distances for Females*, thesis. University of Colorado, Boulder, Colo, 1987.

12. Snook SH, Ciriello VM: Maximum weights and work loads acceptable to female workers. *J Occup Med* 1974; 16:527.

13. Rodgers SH: *Working with Backache*. Fairport, NY: Perinton Press, 1985. (Available through S.H. Rodgers, PhD, 169 Huntington Hills, Rochester, NY 14622.)

14. Putz-Anderson V (ed): *Cumulative Trauma Disorders: A Manual for Musculoskeletal Diseases of the Upper Limbs*. Monograph from NIOSH/DHHS/CDC/DBBS in Cincinnati, New York: Taylor and Francis, 1988.

15. Rodgers SH (ed): Repetitive motions injuries. *Semin Occup Med* 1987;2(1).

# Section 8

# Three Approaches to Specific Ergonomic Problems

*Susan J. Isernhagen*

Ergonomists who approach ergonomics from a physiological base look at the worker and the work with an eye toward physical stresses. The successful ergonomic intervention addresses the interface between the body and potential injury. Because it is the worker who is at that interface, education in ergonomic principles is crucial.

Three approaches to injury reduction through physiological analysis and worker/work/worksite modification are presented in this section:

*Approach No. 1: Prevention and Understanding of Repetition Strain Injury.* Miriam Rowe, an Australian physiotherapist, acted as part of the project team studying repetition strain injury (RSI) of occupations involving the use of keyboards. The resulting work provides a thorough understanding of cause and remediation of this problem, as well as a model of analysis that can be used for problem solving in other diagnostic categories.

*Approach No. 2: Injury Prevention through Ergonomics.* Gunilla Mynerts, a Swedish physiotherapist, has made significant progress in injury reduction at Saab-Scania by involving employees in the ergonomic process. When there is grass-roots involvement, not only is ergonomic information more comprehensive, but workers are encouraged in their responsible attitude toward themselves and their workplace.

*Approach No. 3: Worksite Analysis.* In the United States, worksite analysis is the starting point for industry guidelines in preventing manual handling injuries. Alan Morris, MS, PT, uses not only National Institute for Occupational Safety and Health (NIOSH) guidelines, but also a thorough analysis profile to complete the basic picture of the workplace. This becomes the root of ergonomic design and redesign.

Note: Portions of this chapter were adapted from *Understanding, Prevention and Management of Repetition Strain Injury*, with permission of Department of Occupational Health, Safety and Welfare of Western Australia.

## REPETITION STRAIN INJURY—UNDERSTANDING PREVENTION AND MANAGEMENT[1]

Repetition strain injury is a collective term for a range of injuries to tendon and muscle. It is caused by repeated actions, constrained postures, or both, which produce a cumulative overload of muscles beyond their capacity for immediate recovery. Tenosynovitis, myositis, tendinitis, epicondylitis, and carpal tunnel syndrome may be forms of RSI.

There are reports of RSIs as early as 1713 when a physician named Rammazzini recorded a "disease of clerks and scribes." In 1882, in the prestigious *British Medical Journal,* the existence of telegraphers' cramp was acknowledged. This is considered to be one of the earliest reliable medical records of the existence of RSI.

Similar occupational injuries discussed in medical and trade journals from America, Australia, Great Britain, Japan, and Scandinavia include sewing cramp, scissors cramp, writers' cramp, typists' cramp (in 1920) and rope making hazards (in 1951). Currently these problems have been associated with occupations involving the use of keyboards. A specific prevention program has been designed for keyboard operators to reduce the incidence and severity of RSI.

### Contributing Factors to Injury

A person who has been well trained on the use of the keyboard and who has conditioned his or her muscles should be able to readily perform keyboard tasks with minimal strain. However, the following factors, inherent in the work performed, may override that training and conditioning, resulting in RSI.

#### Use of Excessive Force

Forceful movements may occur several times a day; but, when combined with inefficient use of muscles, they may contribute to an injury. Hitting the keys too hard may be due to:

- inexperience and a belief that forceful movements result in a better job
- lack of training in keyboard techniques
- lack of supervision

---

*Note:* Portions of this chapter were adapted from *Understanding, Prevention and Management of Repetition Strain Injury,* with permission of Department of Occupational Health, Safety and Welfare of Western Australia.

- aggressive working methods
- equipment—for example, keys that require excessive force

*Inefficient Work Habits*

Certain postures may be assumed that are inefficient and cause fatigue. These postures can be due to:

- bad habits
- incorrectly adjusted work stations
- unsuitable work environment
- lack of appropriate furniture or equipment

*Speed of Movements*

When a person performs work at a speed beyond that person's safe level, injury may result. An increase in speed may be the result of:

- incentive systems
- machine-paced work
- high-volume work—for example, too much work to do in too little time
- frequency of movements

### Early Warning—Fatigue

Fatigue is an early warning of the potential for RSI and can manifest locally and systemically.

*Local Fatigue*

There are three groups of symptoms that are associated with local (postural) fatigue:

1. *Biomechanical* (involves muscles, tendons, bone structures): Fatigue in the cervical and thoracic spine leads to an aching, dull pain in the head, neck, and lower back, and between the shoulders; in the affected area of the wrist and hands, tenderness, weakness, swelling, or temperature changes are noted.
2. *Circulatory:* Impaired blood flow results from compression of the main weight-bearing tissues. Symptoms, such as loss of feeling and "pins and needles" are experienced.

3. *Neurophysiological* (involves function of the nervous system): Prolonged backward bending, forward bending, or rotation of the neck may lead to a temporary loss of strength in the arm and hand and a brief loss of ability to perform fine, manipulative skills.

*Systemic Fatigue*

One or more of the following signs and symptoms of systemic fatigue can be experienced in varying degrees:

1. Feelings of fatigue, either over the whole body or in a particular area.
2. Pain—may be described as intermittent or a constant, dull ache; intermittent or sharp shooting; or a burning sensation.
3. Heaviness of a particular area.
4. Loss of sleep, which may cause irritability and anxiety.

In the early stage of injury pain, swelling, or numbness may occur intermittently, whether or not repetitive movement is being performed. If action is not taken, the pain may persist while the operator is performing nonrepetitive movements or even when that part of the body is not being used.

## How To Modify the Workplace

*The Work*

The following interventions directed toward the work may decrease the likelihood of RSI:

- Spreading repetitive tasks among a group of employees to alleviate boredom and varying repetitive tasks to assure that for each employee a variety of muscles are used.
- Rearranging a job to create a whole module of work; for example, giving the operator the whole document, rather than parts of it, to type.
- Designing jobs to incorporate a variety of activities, including breaks from repetitive tasks and periods of work away from the work station.
- Allotting a maximum of five hours of keyboard work per day, including overtime. The work should be divided equally between a morning and an afternoon session, with at least a one-hour break between sessions.
- Limiting continuous keyboard work to 50 minutes out of each hour.

*The Worker and the Work Station*

The following suggestions directed toward positioning the worker and designing the work station may decrease the likelihood of RSI:

- Positioning the worker in optimal alignment to decrease fatigue and concomitant injury:
  — Head should be erect, with the line of vision horizontal with the first line of the draft document or screen characters.
  — Shoulders should be relaxed, upper arms should be by the side of the trunk, and forearms should be level with the A-S-D-F row of the keyboard.
  — Feet should be supported fully, either on the floor or on a footstool.
  — Legroom for unrestricted movement should be provided.
  — Lower back curve should be supported by the backrest of the chair.
  — Wrists should be straight when fingers are resting on the A-S-D-F row of the keyboard.
  — Angle at the elbows should be 90° or greater, with the upper arms parallel to the body.

- Designing the work station to decrease worker fatigue:
  — Work surfaces need to be organized so that all materials, equipment, and controls can be reached easily to avoid continual stretching by the worker.
  — Document holder should be placed at the same distance from the operator as the screen (approximately 500 to 600 mm).

Keyboard operators should consider themselves industrial athletes. The fingers of a word processor operator can walk the equivalent of 17 km in a day. An athlete warms up before running a 3 km run at full pace. Similarly, keyboard operators should warm up at the start of their work period, beginning slowly, and gradually building up to their comfortable typing speed. In this way the body gradually becomes accustomed to the exercise. At the end of the work period, keyboard operators should gradually slow down, allowing the body to return to its preexercise rate.

After being on a vacation or sick leave, the operator should be aware that returning too rapidly to keyboard duties can influence the development of RSI. Time should be allowed for a gradual increase in keying-typing speed to the operator's usual rate.

## ERGONOMIC INTERVENTION AS PART OF AN INJURY PREVENTION PROGRAM

Gunilla Mynerts, an industrial physical therapist, works with employees at the Saab-Scania company in Sweden. She has assisted in the development of a back injury prevention program.

Part of the back school is ergonomic awareness training. Discussions are held with people regarding their workplace. In the structured discussion, chairs, work stations, and work methods are evaluated so the participant will be able to assess better his or her own situation. Good posture is emphasized and methods of accommodating chairs to worktable height are demonstrated. Checklists also are provided to enable the employee to evaluate the worksite from his or her own perspective. Mynerts has commented on some of the results of the ergonomics awareness training.

On one of the items on the checklist, the employee is asked to evaluate the work station, including work height, placement of work materials, type of chair, lighting, and lifting techniques and stresses. Mynerts noted that after this type of ergonomic awareness, workers stated they were able to change their work techniques to some extent (68%) or to a large extent (20%).

Another item on the checklist is designed to initiate ergonomic improvements. As a result, those who observed changes in their worksite felt they were a result of employee initiative. Some employees also received support from the therapists involved.

Mynerts also noted that, in comparison to blue-collar workers, white-collar workers found it much easier to improve their work environment, and had little difficulty recommending the need for, or getting, equipment, as needed. Blue-collar workers experienced some difficulty in bringing about changes, both because of major cost or difficulty in implementing change in areas viewed as large obstacles (eg, design of the machinery) and because of a fear of management disapproval (eg, if they made a change in their worksite and the change was not adopted by all workers). As a result, these workers received more support in consulting with physical therapists at the worksite to effect desired modifications.

In summarizing this experience, Mynerts stated that although the initial intervention is helpful, improvement in the working environment should be monitored on a regular basis so that analysis of working environment and continued ergonomic changes can be made.

## WORKSITE ANALYSIS

In an effort to systematically categorize the worksite for the purpose of injury prevention, worksite analyses have been designed. They assist the evaluator not only in describing and measuring the worksite, but also in indicating potential hazards.

### Lifting Guidelines

An engineering/biomechanical approach has been analyzed and used by Donald B. Chaffin and associates at the University of Michigan, Ann Arbor. Stover Snook of Liberty Mutual Insurance, Boston, Massachusetts, has contributed important research in the areas of work capacity and psychophysical lifting. Much of the work was used by NIOSH in developing an ergonomic manual lifting guide.[2]

Based on the assumption that there is a wide variation in lifting capabilities within the work population, lifting situations can be categorized according to the relative risk of injury that they produce within the workplace. In the *NIOSH Work Practices Guide*[2] two "limits" are described, based on epidemiological, biomechanical, physiological, and psychophysical criteria.

*Maximal Permissible Limit*

The maximal permissible limit (MPL) is defined by the following criteria:

- Musculoskeletal injury and severity rates have been shown to increase significantly when work is performed beyond this limit.
- The limit at which biomechanical compression forces at the L5-S1 disk reach or exceed 1,430 lb (650 kg).
- Metabolic rates would exceed 5 cal/min for most people working at this limit.
- Only 25% of men, and less than 1% of women, have sufficient strength to work above this limit.

*Action Limit*

The action limit (AL) is defined by the following criteria:

- Musculoskeletal injury incidence and severity rates increase moderately in populations exposed to lifting conditions above this limit.

- The limit at which biomechanical compression forces at the L5-S1 disk reach or exceed 770 lb (350 kg).
- Metabolic rates exceed 3.5 cal/min for most people working at this limit.
- Greater than 75% of women, and 99% of men, are capable of lifting loads described by this limit.

**Lifting Situations**

Lifting situations can be categorized in three ways:

1. Above MPL—Unacceptable conditions requiring engineering controls.
2. Between AL and MPL—Acceptable only with engineering and/or administrative controls.
3. Below AL—Represent nominal risk; acceptable for most workers.

Also described in the *NIOSH Work Practices Guide*[2] is a formula for calculating safe lifting limits for lifting based on measuring:

1. Horizontal distance of object to be lifted from ankle midpoint.
2. Vertical travel distance of object to be lifted.
3. Frequency of the lift.
4. Maximum recommended frequency of lift.

By analyzing a lifting task, these four characteristics can be modified to increase maximum load lifted or to affect one variable by changing another. This type of logic is true for most lifting situations. Examples might include:

- diminishing horizontal lift distance to increase load lift safety
- reducing load limit to allow increased repetitions
- reducing vertical lift distance to accommodate heavier loads

**Worksite Evaluation**

To further the analysis of the worksite and incorporate the above ideas, Alan Morris of Dallas, Texas, designed a worksite evaluation that facilitates organized recording of worksite and work parameters to evaluate risk reduction to the worker. (See Exhibit 8-1.) Thorough evaluation of the

**Exhibit 8–1** Worksite Evaluation

---

Company: _____
Department: _____
Function: _____

I. Materials handled _____
   A. Weight _____
   B. Size _____
   C. Shape _____
   D. Stability _____
   E. Rigidity _____
II. Materials handling _____
   A. Surface moved from _____
      1. Height _____
      2. Approachability—distance of feet from load _____
   B. Surface moved to _____
      1. Obstacles, floor fixtures _____
      2. Height _____
      3. Carrying distance _____
      4. Approachability _____
   C. Assistance _____
      1. Mechanical _____
      2. Manpower _____
      3. Availability _____
III. Force application _____
   A. Applying torque _____
   B. Pulling _____
   C. Pushing _____
   D. Rotating _____
IV. Body positioning _____
   A. Standing _____
      1. Degree of bending _____
      2. Static? _____
      3. Moving? _____
      4. Time spent standing _____
      5. Foot support _____
      6. Work height _____
      7. Horizontal distance from work _____
      8. Freedom of movement _____
      9. Overhead obstacles _____

   B. Sitting _____
      1. Time spent sitting _____
      2. Frequency of movement _____
      3. Degree of forward bending _____
      4. Height work surface _____
      5. Chair _____
         a. Adjustable? _____
         b. Arms _____
         c. Low back support _____
         d. Tilt, swivel _____
      6. Foot support _____
      7. Knee clearance _____
   Overhead reaching _____
      1. Height _____
      2. Frequency _____
      3. Stools, ladders _____
V. Work pace _____
   A. Basis for productivity _____
      1. By piece _____
      2. Hourly quota _____
      3. Assembly line/fixed pace _____
      4. Employee control _____
   B. Rest _____
      1. Frequency _____
      2. Duration _____
      3. Sanction _____
   C. Repetitive nature of task _____
      x/hour _____
VI. Work environment _____
   A. Obstacles _____
      1. On floor _____
      2. Overhead _____
      3. Moving _____
   B. Conditions _____
      1. Noise level _____
      2. Lighting _____
      3. Distractions _____

**Exhibit 8-1** (continued)

```
         C. Employee rest areas _____      Comments: _____
            1. Appearance _____           _____
            2. Rest furniture _____       _____
            3. Proximity to workplace _     _____

            4. Sanction for use _____     _____
     VII. Miscellaneous _____             _____
          A. Attitudes _____              _____
             1. Supervisors _____         _____

             2. Employees _____           _____

          B. Other _____                  _____
```

*Source:* Developed by and reprinted with permission of Alan Morris, MS, PT.

physical parameters of the work helps ergonomics practitioners to determine modification strategies. This worksite strategy combines with knowledge of the worker and the work to complete the total picture of injury prevention and maximize productivity.

---

**REFERENCES**

1. *Understanding, Prevention and Management of Repetition Strain Injury.* Department of Occupational Health, Safety and Welfare, The Project Team on Repetitive Strain Injury: West Perth, Western Australia.

2. National Institute for Occupational Safety and Health: *Work Practices Guide for Manual Lifting* (Pub 81-122). Cincinnati, OH: NIOSH, U.S. Public Health Service, 1981.

# PART III

# Pre-employment and Preplacement Screening

Selection of workers based on physical abilities is not new. Historically, size and strength have been very important in selecting the most able workers. With the industrial revolution, larger numbers of workers were needed. Often, the strongest got the job and retained the job. If a worker was not strong enough for the work, injury would most likely occur. This resulted in a work force based on the concept of "survival of the fittest."

When medical professionals became involved in pre-employment screening, it was in the role of establishing a general health level. Employers wanted to hire only people with established health. This was to protect the employees' productivity and to avoid high health care costs both in general health care premiums and in workers' compensation claims. These health-related screenings established absence of significant disease rather than ability levels.

With the current advent of high workers' compensation costs, pre-employment screening is becoming more specific. Advances in methods of establishing criteria for a healthy, low-risk employee indicate that only job-pertinent health information is needed; this would pertain specifically to the stresses of the job. As an extension of this, physical functional testing is receiving strong support to assess minimal acceptable strengths and endurance levels that meet the requirements of the job.

The sections in Part III explore pre-employment screening and preplacement physical examinations. They are:

Section 9     Pre-employment Screening: The Physical Perspective

Section 10    Preplacement Physical Examinations

# Section 9

# Pre-Employment Screening: The Physical Perspective*

*Alan W. Morris and Charles K. Anderson*

Since the first dollar changed hands for a service or product, businesses have come and gone because of changes in the marketplace and economy. Contemporary businesses, however, are faced with a crisis that has the potential of crippling or destroying even the soundest of storefronts, and, eventually, of changing the entire complexion of work in the United States. The issue at hand is the continually increasing drain that workers' compensation losses place on American productivity. The effects are being felt uniformly across the economy and are implicated in what some consider to be one of the most critical periods ever faced in the history of the insurance industry.

The most common source of compensable injury is overexertion, which leads to a variety of conditions, most often sprains and strains of the musculoskeletal system. Of these, injuries to the back and spine are the most frequent, accounting for 20% of all occupational injuries in the United States[1] and an estimated annual cost as high as $30 billion.[2]

Although many sources suggest that 80% of the working population will experience a significant episode of low back pain in their work careers, most of these injuries will not lead to significant or costly workers' compensation claims. In fact, approximately 20% of all reported back injury claims account for greater than 90% of the costs associated with low back pain.[1,3]

What, then, is the prudent employer to do to protect his or her workers and assets from the hardships of costly, drawn out back injury claims? An ergonomic approach to the design of the worksite and job tasks is recommended first to minimize exposure of the work force to physical and psychological stressors.

---

*Portions of this chapter were adapted with permission from *Occupational Health & Safety* (December 1985; 55 [12]), Copyright © 1985, Stevens Publishing Corporation.

The National Institute for Occupational Safety and Health (NIOSH) has studied carefully manual materials handling and summarized its recommendations in the *Work Practices Guide for Manual Lifting*.[4] In 1986 the Department of Labor began exercising compliance controls for those companies with an incidence of reportable back injuries above acceptable standards for the industry. Citations were to be given to companies with a high incidence of back injuries and that failed to attempt to find administrative or engineering solutions to the problem.

Safety and training programs designed to make the employee aware of his or her responsibility for safe behaviors and performance are typical examples of administrative solutions that have been useful in controlling the incidence and costs of injuries. Selection of workers is another administrative control recommended by NIOSH. Selection is appropriate for jobs that *require* moderate to high levels of identifiable and measurable degrees of physical performance as necessary tasks. The intent is to identify people who possess the capabilities of meeting or exceeding the requirements of the job. Selection through medical screening is basically a tool for "protecting susceptible workers and preventing the economic consequences of job-related injuries and illnesses."[5] Screening people for work involves an array of complicated social, economic, moral, and legal issues that will impact on the ultimate useability of nearly all screening and selection procedures.

## EVALUATION OF ALTERNATIVE SELECTION PROCEDURES

Selection procedures vary widely and are based on the size and resources of a company and the attitude of its management. In evaluating the useability of any testing procedure to measure physical performance, the procedure must satisfy a variety of medical, social, economic, and legal criteria, as described by Chaffin[6,7] and outlined in the NIOSH *Work Practices Guide*.[4] There are five basic evaluation criteria: (1) safety of administration, (2) reliability, (3) job relatedness, (4) practicality, and (5) predictiveness.

### Safety of Administration

Tests of physical performance, of course, should be safe to administer. The potential risks and exposure to the person should not outweigh the value of data to be gained. Likewise, the test procedure should not set specific goals for the worker or encourage the worker in any way to perform beyond a level that feels safe and comfortable to him or her. Of course, these tests should be administered by trained people who can recognize and respond to physical complaints during testing. It is recommended that subjects be screened medically beforehand.

## Reliability

The second consideration in evaluating the useability of a test is the reliability of the data provided. There is great variation among tests in providing reliable, measurable, and reproducible results. For instance, for isometric strength testing,[7] the percentage of deviation expressed as the test/retest error, or "coefficient of variation" is in a range of 10% to 15%; whereas, for subjective assessments of capability, the coefficient of variation can be much higher.

## Job Relatedness

To assess the applicant's ability to perform the physical tasks of a given job, the test must be job related. This aspect of screening is critical for compliance with the Equal Employment Opportunity Commission (EEOC) guidelines and the prevention of discriminatory action suits. For example, administration of intelligence tests would not be considered an appropriate way to assess a person's ability to lift 100-lb sacks.

## Practicality

The usefulness of a selection procedure will be based, in part, on the practicality of its development and implementation. The employer will need to consider the costs of a job analysis, protocol design, incumbent testing, and the necessary validation studies as up-front expenses. The equipment to be used must be reasonably priced and should easily accommodate different simulated job tasks. Testing procedures should be quick and simple to deliver, and the training of people who will administer the test should not require excessive time or personnel requirements.

## Predictiveness

Finally, and most importantly, the selection procedure should predict risk of future illness or injury and/or performance. This phase requires a careful comparison of test results to injury and performance data obtained from the tested population. It is a necessary requirement set forth by the Department of Labor in the *Uniform Guidelines on Employee Selection Procedures* as part of the Civil Rights Act of 1978.

## ALTERNATIVE FORMS OF SCREENING

Those five criteria can be used to evaluate the typical approaches to screening that are in use today. The current trends in physical screening of workers range from evaluating the person on the basis of individual characteristics to obtaining medical histories and physical examinations, radiographs, and strength tests, to name a few. The testing of strength, however, shows the most promise from a variety of perspectives.

### Individual Characteristics

Although the work by Rowe,[1] Frymoyer,[2] Snook,[3] and others indicates that low back pain occurs most commonly between the ages of 35 and 45 and that men and women are affected equally (Nachemson), there is no strong correlation among weight, height, stature, or the Davenport Index and the incidence of low back pain. The degree of lumbar lordosis and the presence of minor scoliosis do not seem to increase the likelihood of low back pain. Most structural abnormalities that are seen on radiographs are not related to back pain incidence, with the *possible* but questionable exception of spondylolisthesis. There are reports, on the other hand, that suggest factors, such as physical fitness, abdominal muscle strength, trunk muscle balance, exposure to heavy lifting or vibration, and even cigarette smoking, are positively correlated to episodes of low back pain.[2,3]

Although the entirety of data does not indicate sufficient sound evidence on which fair hiring decisions can be made, one undisputed predictor of future risk is a past history of low back pain. People who have experienced significant low back pain in the past are at four times the risk of future episodes compared to those who have no prior history.

## Medical History and Physical Examination

Selecting and maintaining a work force through the application of medical criteria most often involves the assessment of current health status and a medical determination of the workers' future health risks. The latter effort is classified as predictive screening—an attempt by industry to identify the people who are most likely to become sick or injured before they become an economic liability. Strong financial incentives spawned what could be called a preemployment "witch hunt" to identify and select out the person with an assumed low back disorder. The caveat is summarized by a railroad executive who stated, "In the face of staggering claims and settlements, we must do everything possible to exclude the serious back injury."[8] Fighting for financial survival, the industrial community began applying broad-based hiring policies to screen out abnormalities of the back and spine with seemingly little regard for fairness, accuracy, or the applicant's actual ability to safely perform the job in question.

Medical examinations of some form are widely used for determining an applicant's current health status and susceptibility to certain environmental exposures. Examinations of vision and hearing and a gross examination of the cardiopulmonary system are fairly routine. When physical or environmental hazards warrant, a thorough clinical examination of applicants by a physician is considered to be the first step in a thorough screening program. In many cases, however, the correlation between physical findings and the worker's current ability to perform job tasks is based on subjective impressions rather than objective data. The interpretation and use of this information in the hiring process also are quite variable, and the reliability and accuracy of the information provided can be questioned.

The physical examination is designed to identify clinical signs of spinal dysfunction that may indicate future risk. Although posture and range of motion are not reliable predictors in and of themselves, abnormalities found on a physical examination may indicate a need for more detailed evaluation. If nothing else, it provides an opportunity to screen for surgical scars and artifacts of invasive procedures.

Because of the high recurrence rate of back pain episodes, the history of previous incidents is an important part of identifying high-risk applicants and directing them toward prevention efforts and suitable job placement. Also, it appears that the severity and pattern of onset of previous episodes have predictive value. Lloyd and Troup[9] identified four factors in the injured worker that seem to be predictive of recurrence: Residual leg pain, an injury associated with a fall, lost time in excess of five weeks, and a history of two or more prior episodes increased the likelihood of recurrence as the number of these factors increased.

## X-Rays

Pre-employment X-rays of the lumbar spine became a popular mechanism for categorizing applicants as either acceptable or unacceptable for hire, in spite of their very low predictive value in identifying future back disability. But, still, the low back X-ray is one of the most prevalent forms of screening used to assess the risk of back injury, with an estimated one million preemployment films being taken each year.

Many sources within the literature expound on the very low benefit-to-risk ratio of the pre-employment low back X-ray[1-3,8,10-13] and the general consensus is that their use as a routine screening tool should be abandoned. Support for this posture is argued convincingly by Rockey and Fantel,[8] who cite low sensitivity (the ability of a test to identify correctly people with a condition) and low specificity (the ability of a test to correctly identify people free of a condition). These indicators lead to a very low predictive value for X-rays as a way of identifying people who are at increased risk of experiencing a disabling low back injury.

Another issue to consider in the use of routine pre-employment X-rays is the unnecessary health risk associated with the procedure. Low back X-rays are the largest single contributor to gonodal irradiation in the United States,[14] which could increase the risk of diseases caused by irradiation by as much as 25% (National Research Council [US], 1972).[15]

The possibilities of exposing a woman with an unsuspected pregnancy to unnecessary X-rays are calculable, a position which can increase the risk of childhood leukemia between twofold and tenfold.[16] Pope concludes that from the standpoint of health risks, "low back X-rays should always be considered a potential health risk and any unnecessary use should be discouraged."[14]

Hiring decisions that are based solely on spinal X-ray do not meet the requirements set forth by the EEOC. X-ray screening fails to meet the standards of "job relatedness" and potentially discriminates against those with a perceived handicap as defined in the Rehabilitation Act of 1973. Under Section 504 of this Act, asymptomatic people denied employment on this basis may have legal recourse at state and federal levels.[8] One such decision was upheld in the precedent-setting case of *OFCCP v E E Black, Ltd.*[17]

## Strength Tests

A person can be evaluated from the standpoint of physical performance capabilities. Muscle strength measurements and the physiological (metabolic) response to workloads can be objectively measured and documented for comparison to job requirements.

Strength testing from the authors' point of view can be divided into tests of functional strength (ie, the ability to exert lifting, pushing or pulling forces) and tests of "muscular strength," in which the strength and balance of muscle groups are examined through a range of motion. Functional strength tests typically involve having the person use the body in whatever way he or she chooses to accomplish the requested exertion, such as applying a lifting force to a set of handles. Muscle strength tests, on the other hand, involve isolating muscles or muscle groups and examining their strength potential throughout their physiological range of movement, as in testing the trunk flexors or extensors.

*Functional Strength Tests*

Clearly, testing for functional strength is the most job-related form of strength testing. The positions encountered during handling activities can be closely simulated. Also, a functional strength test is a better evaluation than a "muscle strength" test of the person's ability to do a task, in that it allows the person to compensate by using different postures and/or synergistic muscle groups if one particular group is dysfunctional or weak. A third advantage of functional strength testing is that the test administrator can see how the person uses his or her body in the exertion of force (ie, the person's body mechanics). The authors recommend that test subjects not be instructed in new lifting techniques as a part of the test procedure. The disadvantage of functional strength testing is related to the nature of the test, which makes it difficult to isolate the areas of weakness to particular muscle groups or structures and thus develop a rehabilitation program for the person.

There are basically three types of functional strength tests that measure a person's ability to exert forces in functional situations: static (isometric), dynamic, and psychophysical tests.

*Isometric Testing.* As described by Chaffin,[6,7] isometric strength testing involves the exertion of a force in a given posture against a fixed resistance. The force exerted is objectively measured by a load cell and is compared to the force requirements of the job. Lifting height and horizontal reach distance dictated by the job requirements can be reproduced in the test protocol. With simple modifications, the test can be accommodated to simulate pushing and pulling activities as well.

Validation studies of this protocol indicate that workers who were placed in jobs where job demand exceeded isometric strength capabilities had incidence rates of sprains and strains three times higher than those whose isometric test strengths exceeded job strength requirements. The test also is considered safe, reproducible, and simple to administer,[7] and is recognized in the *NIOSH Work Practices Guide for Manual Lifting*.

In spite of these promising results, isometric testing has been criticized for several reasons. First, isometric lifts traditionally test only in one geometric plane of movement whereas normal human function normally cuts across three planes. Second, it tests strength in only one position devoid of joint movement. This does not allow for the dynamic changes in posture, joint loading, and muscle length/tension ratios that accompany functional activity, nor does it account for acceleration and deceleration of body segments that make up 82% to 84% of body movements (D. R. McIntyre and J. A. Sawhill, unpublished data, May 1984).

Currently, it appears that isometric testing offers a validated, documented, and potentially predictive screening tool, which also serves as an excellent starting point for research and comparison of other test protocols.

*Dynamic Tests.* Whereas isometric tests measure the forces produced in a fixed position (when velocity = 0), dynamic testing (isokinetic or isodynamic) allows the measurement of forces throughout a range of movement. As in isometric testing, task variables, such as vertical height at the beginning and end of the lift, horizontal reach distance, foot placement, and lifting speed, can be incorporated into the test. In an isokinetic test the velocity of movement stays constant at a preset but adjustable rate, and resistance accommodates to match the force applied. Although some people argue that the constant velocity of isokinetic testing does not realistically simulate functional activity, proponents argue that it measures actual force output capability at every point throughout the lifting arc.

Isodynamic tests, which became available in the early 1980s, offer yet another choice. With this type of testing the velocity varies whereas the resistance or load is preset. This even more closely imitates actual lifting conditions than a fixed-velocity test.

Isokinetic lifting test data on normal subjects and back injured patients have been collected by Kishino and colleagues.[18] Comparisons of isokinetic and isometric tests indicate consistencies and reproducible relationships between the two tests, as well as predictable patterns among back injured people. Those who intentionally perform at submaximal levels also are easily identified, a factor that may have implications in identifying symptom magnifiers among back claimants.

These studies indicate that isokinetic lifting tests also are safe and reliable, although equipment cost may be an issue relating to practicality of delivery in the average industrial setting. A second issue of concern is job relatedness. The equipment available to date varies greatly in the degree to which it accommodates lifting movements (and push/pull tasks) outside of one plane.

*Psychophysical Tests.* Psychophysical testing, as described by Snook and others, also provides reliable, reproducible data regarding a subject's tolerance for specific tasks. This approach is based on the concept that human capabilities are determined by a combination of physical, perceptual, and judgmental factors that will influence the person's performance.

Psychophysical tests are dynamic tests in which the subject is asked to perform a simulated job task by handling a box in a way that simulates the activity of the job. The job can be very closely replicated by adjusting variables, such as the direction of movement (eg, lift, push, pull), frequency of movement, and distance moved. The person is allowed to adjust the amount of weight in the box until it is at a level he or she would be willing to work with for an entire shift. The weight that subjectively feels safe and comfortable for that task is considered the safe upper limit for that person.

Although the safe upper limit weight for a person is based on his or her subjective report, Snook[19] indicates that test/retest values can be expected to vary within 15%. Chaffin and Anderson suggest that this method "may be the most accurate method of determining a person's acceptable performance limit for a given task."[20]

Unfortunately, psychophysical testing is a somewhat time consuming and personnel intensive protocol and requires maximal cooperation from the subject. Also, there are indications that people overestimate their eight-hour capacity during a test that may last only 20 minutes. This tendency can become particularly acute when a job candidate realizes that his or her self-elected maximum weight will be used to evaluate employment opportunity.

## Muscle Strength Tests

Muscle strength testing deals with the evaluation of the strength and function of trunk muscles, namely the flexor, extensor, and rotary groups. The ability of muscles to produce forward bending, backward bending, and rotary movements of the trunk can be measured in several ways.

Isokinetic testing of the trunk muscles began in the mid-1970s. In testing of normal subjects by Casey Lee, Davies, Gould, Mayer, and others, reproducible values of trunk muscle function have been established. Strength to body weight, rate of muscle contraction, and flexor/extensor ratios have been developed and appear to be consistent across a broad population of normal subjects. Interestingly, people with back dysfunction will deviate from these normal values in predictable, reproducible ways. Research indicates that abnormal trunk flexor/extensor ratios exist in some asymptomatic people and, in fact, may be present before the onset of symptoms. The potential usefulness of these tests as predictors of future injury is therefore encouraging (T. Mayer, personal communication).

Baseline information obtained during preemployment assessment with this equipment also may be useful to quantify the degree of limitation after an injury and/or during the rehabilitation process. Because of the sophistication of the documentation obtained from isokinetic trunk testing, submaximal efforts can be identified by test/retest inconsistencies. This may give additional objective insight into the legitimacy of an alleged injury and its resultant disability.

Although considerable data have been developed, dynamic testing of the trunk muscles needs further research regarding specificity, sensitivity, and predictive value in terms of future risk of injury. Moreover, it is difficult for a test of this nature to be directly related to job tasks. Hence, job relatedness is more difficult to support with muscle testing than with functional testing.

Another obstacle in the industrial application of this equipment is that its high cost is difficult for most companies to cost justify. In most cases, this equipment is most attractive to clinics, hospitals, and testing centers.

## DEVELOPMENT AND IMPLEMENTATION OF A SCREENING PROGRAM

The development and implementation of a screening program need to be approached carefully to assure that the final program is truly effective and meets legal standards outlined by EEOC and the Rehabilitation Act of 1973. There are four basic steps that are critical to the implementation of a prudent screening program: (1) identification of high-risk jobs, (2) job analysis, (3) test design, and (4) validation.

### Identification of High-Risk Jobs

A clear understanding of the trends in musculoskeletal injuries and their causes must be developed to identify high-risk jobs. These data can be gained from accident reports, insurance loss runs, the Occupational Safety and Health Administration's log of reported injuries, and infirmary visits. By comparing these data with worker exposure hours, high-risk target jobs can be identified for detailed job analysis.

## Job Analysis

The EEOC has set clear guidelines regarding the use of specific testing procedures. To be considered legal and nondiscriminatory tests must be clearly job related and must actually be able to predict the applicant's ability to perform safely the necessary tasks within the job in question.

The ability to structure a test that is related to work activity and to effectively match a worker's capabilities with job demands depends on a careful job analysis. High-risk jobs must be dissected to identify the most stressful tasks involved. The weight of materials handled, push/pull forces required, and postures involved must be objectively measured and must be documented. Repetition of activity, task duration, and peripheral variables, such as exposure to heat, vibration, or trip hazards, are important as well. Cardiovascular stress and fatigue also are considerations in musculoskeletal injuries.

The job analysis information can be used to guide efforts at developing engineering controls. If reasonable attempts fail to bring the task within the capabilities of the majority of the work force, the employer must show that the specific task is critical to the satisfactory performance of the job in question and is of "business necessity" before it can be included in the screening protocol. A thorough job analysis is of particular importance if the test unintentionally discriminates against a protected group, causing "adverse impact." For example, women generally score lower on strength tests than men and thus would be impacted more adversely. Obviously, it is desirable therefore that the test evaluate the ability to perform tasks that cannot be changed or eliminated from the job in question.

## Test Design

Job analysis provides the information necessary to design tests that are job related and predictive of a person's ability to safely perform the required tasks. In the design process, consideration should be given to routine, as well as occasional or infrequent, tasks required of the worker. For entry-level positions, for which promotion or transfer to alternative jobs with potentially different tasks is likely, the test design should be reflective of those tasks as well. For job tasks that require specific postures dictated by the vertical and horizontal location or by a limited approach to the work, the test design should simulate the position in which forces are actually applied. The determination of qualifying scores is based on the lifting requirements, push/pull forces, and cardiovascular stress documented in the job analysis.

## Validation

Validation of the screening protocol is an important step in determining the tests' predictive value, and is a requirement mandated by the Department of Labor. It must be shown either that people with a test performance score below the cutoff score are at a significantly higher ($p = 0.05$) risk of injury or illness than people with scores above the cutoff or that test performance in some way is positively correlated with the person's current ability to safely perform the tasks required by the job.

The guidelines set forth by the Department of Labor discuss various validation procedures. The most defensible form of validation is a prospective study. This type of study involves testing a group of newly hired workers but not making any placement decisions from the test results during the validation phase. In essence, these newly hired workers are treated as they would have been under the current hiring procedure. Their performance, injury experience, days lost, workers' compensation costs, and tenure are then monitored into the future until sufficient data have been gathered to statistically substantiate (at the $p = 0.05$ level of significance) that candidates scoring above the cutoff criteria are at lower risk of injury and/or have better performance on the job.

A retrospective validation study can be done with incumbents to assess more rapidly the viability of a screening battery, but a retrospective study does not give evidence regarding the ability of the test to *predict* future risk of injury or performance. In a retrospective study, a group of incumbents are tested and then their historical performance and injury data are gathered. Data from incumbents with scores above the criteria are compared with data from incumbents with scores below the criteria to assess whether there is a significant difference in the risk of injury or performance. The advantage of a retrospective study is that it can be completed quickly, because there is no need for a time delay between testing and injury/performance data collection. As mentioned above, though, the study does not indicate whether the test battery is predictive of future risk of injury. Therefore, it is important that a prospective study be started at some point before the battery is put into use for selection purposes.

## CONCLUSION

Clearly, the area of pre-employment selection of workers is in a state of evolution away from the guesswork and "crystal ball" approach and toward more objective, predictive, and fair testing protocols. There is promise, especially in the areas of dynamic, isometric, and psychophysical strength testing. Most of these tests do require further validation and scrutiny with regard to sensitivity, specificity, and predictive value before they can be considered legally supportable and really useful in preventing injuries. The task of testing and developing these new protocols lies with responsible companies that are willing to work with the medical community to abandon archaic testing formats and to confront the frontier.

Finally, preplacement selection procedures should not be considered as the sole means to reduce the number of overexertion injuries. The implementation of a screening program does not exonerate the employer from the responsibility of providing a safe, well-designed workplace, or from making reasonable accommodations and prudent placement decisions for high-risk applicants.

## REFERENCES

1. Rowe ML: *Backache at Work*. Perrington Press, 1983, p 6.
2. Frymoyer J: The challenge of the lumbar spine, abstracted. Minneapolis, Minn, 1985.
3. Snook S: Proceedings of the conference on "Industrial Low Back Pain", University of Vermont, October 1983.
4. US Department of Health and Human Services: *NIOSH Work Practices Guide for Manual Lifting*. Cincinnati, Ohio, 1981.
5. Rothstein M: *The Medical Screening of Workers*. Washington, DC, Bureau of National Affairs, 1984.
6. Chaffin DB: Erogonomics guide for the assessment of human static strength. *Am Ind Hyg Assoc J* 1975:505.
7. Chaffin DB: Functional assessment for heavy physical labor. *Occup Health Saf* 1981 (January).
8. Rockey PH, Fantel JD: Discriminatory aspects of pre-employment screening: Low back x-ray examinations in the railroad industry. *Am J Law Med* 1979; 5:197-214.
9. Lloyd DCRF, Troup JDG: Recurrent back pain and its prediction. *J Soc Occup Med* 1983; 33:66-74.
10. Frymoyer JW: Proceedings of "Industrial Low Back Pain in Industry Conference", University of Vermont, October 1983.
11. Montgomery CH: Pre-employment back x rays. *J Occup Med* 1986; 18:495-498.
12. Splitahoff CA: Lumbosacral junction: Roentgenographic comparison of patients with and without backaches. *JAMA* 1953; 19:152.
13. LaRocca H, Macnab I: Value of pre-employment radiographic assessment of the lumbar spine. *Ind Med Surg* 1970; 39:253 .
14. Pope MH et al: *Occupational Low Back Pain*. New York, Praeger Publishing, 1984, pp 222-225.
15. National Research Council Advisory Committee on Biological Effects of Ionizing Radiation. The effects on populations of exposure to low levels of ionizing radiation. *Proc. Natl Acad Sci USA* 1972, p 55.

16. Bross IDJ, Nataragan N: Leukemia from low level radiation: Identification of susceptible children. *N Engl J Med* 1972; 287:107–110.

17. *OFFCP v E E Black, Ltd,* Dept. of Labor Case No. 77-OFCCP-7-R, 19 FEP Cases 1624 (1979).

18. Kishino N, Mayer T, Gatcher R, et al: Quantification of lumbar function. IV. Isometric and isokinetic lifting simulation in normal subjects and low back dysfunction patients. *Spine* 1985; 10:921–927.

19. Snook S: The design of manual handling tasks. *Ergonomics* 1978; 21:963–985.

20. Chaffin DB, Anderson G: *Occupational Biomechanics*. New York, John Wiley & Sons, 1984.

# Section 10

# Preplacement Physical Examinations

*William P. Fleeson*

Preplacement medical examination is an appropriate method to evaluate a potential employee's physical suitability for a job. An appropriately designed preplacement physical examination can be made more thorough and useful by including screening components performed by physical therapy professionals, supplemental to the medical examination.

Any company that has jobs that carry the risk of injury should require some form of medical examination of an employee at the time of hiring. Management should make sure this examination is performed before the employee starts work.

Traditionally, such physical examinations were called "pre-employment" examinations and the employer hired the applicant only if he or she "passed the physical." Those days are gone as a result of guidelines outlined by the Equal Employment Opportunity Commission, which eliminate discrimination against handicapped applicants. Now an applicant must be hired *before* being examined medically, and the procedure is described as a "preplacement" examination.

Nevertheless, it is still important to evaluate a new employee medically to ensure that he or she is (1) physically suited to perform the job and (2) not at significant risk of illness or injury to himself or herself or others when working. In addition, once the decision has been made to hire a person, the medical examination will assist in placing that person in a position appropriate for his or her physical condition.

It is imperative that this type of screening examination be standardized. It must evaluate only job-related requirements, and it must be performed equally on all applicants. Only if an applicant is physically unsuited for the job and/or the employer cannot reasonably modify the position can employment of the applicant be declined.

A new concept in preplacement examinations has been used at the Industrial Medicine Center in Duluth, Minnesota, to assess an applicant's physical and health status and to provide managers with factual medical data that they can use to make management decisions. The Directed Occupational Medicine Examination© (DOME) rectifies the shortcomings of traditional medical screening and examination procedures used in the hiring period.

## PROBLEMS WITH TRADITIONAL MEDICAL EXAMINATIONS

A medical examination, conducted in the traditional manner with the traditional data format, can be a source of unnecessary costs, misleading results, and even legal liability. In the past, management usually relied on contracted physicians to do "a complete medical" on applicants. The basic heart, lung, and blood pressure examination was expected to "pick up any problems" in a potential employee.

On analysis, however, the standard examination was originally designed as an individual disease and symptom checklist for hospitalized or clinic patients that were sick. It was not designed to screen for physical capabilities, or for risks of injury in a certain job. The format of the traditional examination, with laboratory tests, x-rays, and so forth, did provide a great deal of medical information, but the information was generally not specific or relevant to management's basic need to know about the applicant's actual job suitability.

Managers should not rely on medical input that is not matched to a job-related evaluation. Using such data in making decisions to hire or not to hire could easily put management into unwitting violation of antidiscrimination law.

The traditional examinations are still being performed by the thousands. Most traditional physicians would be reluctant to alter their old standby format (history and physical examination with laboratory tests and x-rays), that has been performed for decades. But progressive managements familiar with occupational medicine concepts are asking for preplacement examinations that can provide the specific information to answer their basic question, "Is this applicant suited for this job?"

## THE DIRECTED OCCUPATIONAL MEDICINE EXAMINATION

Preplacement examination can be successful. It can provide the information to answer management's specific questions about suitability for work and it can be job specific. Such a preplacement examination must be structured so that it allows the occupational medicine physician (and thereby the manager) (1) to assess the individual applicant versus the specific requirements for his or her job; then, (2) to make a specific statement of risk; and (3) to treat all applicants equally, using not only individual considerations but also professional medical judgment.

A directed preplacement examination takes the best from the traditional medical examination, discards irrelevant or unnecessary parts, and adds new components to the examination procedure. The result provides management with an accurate and fair tool for assessing the risk of injury in applicants. The examination consists of four parts.

First, the applicant provides a thorough medical history. The questions are designed to uncover past or existing medical conditions that would have a bearing on the applicant's ability to perform the job. For example, history of fractures or deformities of the feet would be important for an applicant required to climb ladders every day.

Second, a medical examination is performed. It includes the areas addressed in traditional medical examinations (eg, measuring vital signs; assessing vision and hearing, heart and lung status). The examination also focuses on a musculoskeletal evaluation, particularly of the spine and extremities to assess for flexibility and muscle tone and balance. Screening includes assessing for neurological problems, as well as attention factors, the applicant's response to instructions, and his or her attitude.

Third, laboratory and ancillary testing often is required. Some employers require urine drug screening; others require certain X-rays or other tests. For example, a Mantoux test for tuberculosis is required for nurse assistants; pulmonary function testing and a chest film are required for asbestos workers. Testing often is mandated by Occupational Safety and Health Administration (OSHA) guidelines; at other times particular laboratory testing is requested specifically by an employer.

Fourth, job simulation testing is performed. The author has developed tests to simulate the jobs for which applicants apply. Simulations are based on physicians' observations and measurements of actual jobsites over several years and on knowledge of the type of work the applicant will be performing. The test is administered by a physical therapist and allows for further evaluation of an applicant's physical suitability for a specific job.

The simulations measure the applicant's general fitness and agility, strength, and ability to perform specific job-related tasks. The physical therapist is trained to observe specifically for conditions that predispose to injury. Also, throughout the testing, the physical therapist evaluates and rates body mechanics.

As an example of the examination process, an applicant applying for a nurse's assistant position will first complete the history, medical examination, and laboratory testing. If there are no apparent risks to further testing, the applicant participates in the job simulation. Squatting, kneeling, and bending as well as aerobic capacity and patient transfer procedures are some of the items included in the testing. Some tests, such as grasp strength, are measured and compared to norms.

After all four parts of the directed occupational medicine screening are completed, the physician considers the results of all components taken together and weighs them against the applicant's job requirements. Based on all the data, the physician can make an estimate of risk of injury and a judgment of suitability for the job. The physican's recommendation to the employer then synthesizes relevant aspects of the applicant's medical history and health status, body mechanics and ability, and laboratory results.

The physician can make specific recommendations regarding whether the applicant is at low, average, or high risk of injuring himself or herself or others in the position, and can recommend modifications to the job or specific changes in the applicant's status. For example, the examination may indicate that the applicant has good strength but poor body mechanics and needs further training. Or, the applicant may have good body mechanics and strength, but has a previously unsuspected heart murmur, high blood pressure reading, or active ulcer disease that requires investigation by a private physician. In such a case, the physician would recommend provisional hiring or deferring a decision.

The DOME has been well received by both applicants and employers. In general applicants not only realize they are being evaluated objectively but also develop further self-confidence from having performed a task easily. Others are surprised by deficiencies in their training or strength. There have been only a few cases in which it was difficult for the physician to make determinations of risk. The examinations have enabled management to make more informed decisions on the applicant's suitability for the available job.

## OTHER REASONS FOR DOING PREPLACEMENT PHYSICAL EXAMINATIONS

Besides assessing physical suitability for a job and risk of injury there may be other reasons for performing a preplacement examination. These are often dictated by specific situations. Generally these apply to the medical portion of the examination, and are requested by the company on the basis of tradition or on special needs of the workplace. As examples, preplacement examinations may be of value in screening for current illness (eg, hypertension or cardiac disease) or for predisposing conditions (eg, spondylolisthesis, spondylolysis, or rheumatoid arthritis). Examinations may be used to prevent future injury or illness (as in cholesterol screening) or to detect communicable diseases (eg, tuberculosis, syphilis, salmonellosis). Some examinations are designed to uncover potential major medical risks (eg, future cardiac or pulmonary disease), and some attempt to evaluate mental or psychological predisposition to future health problems (eg, bulimia). In other instances companies require screening for problems specific to their industry (eg, respiratory allergy screening in certain metal industries; screening for sinus problems in dust industries).

All of the above can at times be appropriate uses for preplacement examinations. However, preplacement examinations are *not* suitable for use if management plans to use them as the basis for hiring or not hiring, "passing or failing" the examination, or uncovering confidential information as a substitute for personnel screening; or as a mechanism to predict exactly future health problems. Preplacement examinations must not be used as a basis for discrimination, and both employer and employee should realize that a preplacement examination is not a health checkup or a complete physical. It is an examination done only to determine risk and suitability for a job.

Any potentially discriminatory testing procedure should be eliminated from the evaluations. Preplacement screens cannot discriminate against an applicant on the basis of arbitrary or nonobjective criteria. The only justification for declining to hire an applicant is based on the applicant's ability or lack of ability to perform the job.

The relevant medical and legal literature indicates that employers and medical and physical medicine professionals must keep in mind that:

1. Employment denial is justified if the person has a reasonable chance of injuring himself or herself or others, or of failing to do the job safely.
2. All applicants must be evaluated with a similar format and procedure; however: (a) Examinations also must be individualized, based on the specific job in question and on the individual's specific health status; and (b) employment screening must relate to the ability to do the specific job only.

An employer may decline to hire based on only bona fide occupational qualifications (BFOQs) that are unequivocal. Such BFOQs usually are related to safety considerations, especially safety to fellow employees. Any potentially discriminatory test must be predictive of, or significantly correlated with, important elements of work behavior that make up the job being applied for.

Any deviation from these principles is unfair, discriminatory, and should not be tolerated. Performing a directed preplacement screening examination as described above is the only justifiable and legally defensible way to evaluate and choose among potential job applicants.

## SUCCESSFUL SCREENING EXPERIENCE

The directed occupational medicine evaluation has worked very well in the evaluation of people's level of risk in the job. It also has been valuable to employers in helping them determine if applicants are suited for the job for which they have applied.

It is as important to tell an applicant that he or she is not suited for a job as it is to make this information known to the employer. An occupational medicine physician is not doing a job applicant any favor by allowing him or her to report for work in a position requiring greater physical strength than the applicant possesses. Such a mistake is both costly to the employer and risky to the applicant's health. For example, it would be inappropriate and dangerous to ignore vertigo in a person applying for a position as firefighter. It would be foolhardly to allow a person with 30 lb lifting capacity to work as a stonemason, and it would be unwise to hire a person with chronic obstructive pulmonary disease in coal mining.

In using these special preplacement examinations, the short- and long-term results have been to reduce injuries. When employers have accepted the verdict of high or low risk and placed applicants accordingly, the result has been fewer injuries. When employers have ignored the risk assessment, more injuries have resulted. In general, employers have reported consistently that the frequency and overall cost of injuries have decreased as a result of these specific preplacement occupational medicine examinations.

# PART IV

# Care after Injury

Providing immediate care for injuries that occur on the job is one of the biggest challenges for both industry and medicine. Most professionals agree with the principle, but the practical application of an immediate delivery system has been more difficult to accomplish. However, reduction in symptoms, early return to work, and prevention of chronic conditions are worthwhile goals.

The sections in Part IV address three aspects of efficient care after injury. Information on traditional treatment techniques, which is available elsewhere, is not included.

### Section 11   Immediate Care Delivery Systems

Comprehensive medical systems that interface with industry to provide efficient immediate care are discussed. Examples of such systems provide the reader with issues and solutions in design of effective programs. Communication, time efficiency, simplicity, and a return-to-work function philosophy are major strengths.

### Section 12   Role of the Physical Therapist in Early Intervention

With a focus on care of musculoskeletal injury, experienced physical therapists present philosophical and practical overviews of early intervention. The approach emphasizes the whole worker and functional capability.

## Section 13   Stabilization Principle for Physical Management of Work Injuries

Treating orthopedic injuries can lead to a focus on the injured part rather than the whole person. Movement is traditionally emphasized and, as a result, stabilization may be overlooked. This segment returns to the basics with rehabilitation principles that allow the back to become a body stabilizer, both returning function and preventing reinjury.

# Section 11

# Immediate Care Delivery Systems

*Susan J. Isernhagen*

### RECOGNITION OF EARLY CARE

Three concepts related to care after injury have changed work injury care delivery systems:

1. The longer a worker is off duty because of an injury, the less the chance for a successful return to work.
2. Rest, a conservative method of allowing healing, is a cause of deconditioning of injured workers.
3. Powerful psychological factors affect the outcome of work-injured patients when return to work is not immediate.

When the goal of work-injury treatment is more than healing, progressive rather than conservative concepts are used. For example, in sports medicine, an athlete's goal is to return to the game as soon as possible. An injury will be treated to allow maximum function while the injured part is protected. In this way, healing can take place and further reinjury can be prevented; yet, even strong athletic activity may not be precluded. In geriatrics it is being realized that inactivity is a potential killer and that activity after fracture or injury will assist in preventing the side effects of rest.

Immediate care of work injury has also been affected by the cost of workers' compensation claims. It is the long-term medical costs and disability settlements of those chronically ill or in delayed recovery that have been targeted.

The resultant change in appropriate immediate care has been the elevation of the return-to-work goal to a level equal with the healing goal. Several models of medical delivery systems to meet the immediate needs of the injured worker, the employer, and the medical practitioners have been developed.

## COMPONENTS OF A DELIVERY SYSTEM FOR CARE AFTER INJURY

In most occupational medicine programs that emphasize both early return to work and healing there are five principles applied to care after injury. They reflect not only a change in philosophy from conservative to active, but also the positive qualities of other care delivery systems. Each is highlighted below.

### Immediate Response

In the successful occupational medicine program forces are mobilized for quick response to the injured worker. Care may be for minor injuries, such as cuts and scrapes, or for any level of more severe trauma. The triage principle, exercised by a coordinating nurse or physician, facilitates early intervention, with referral to levels of care appropriate to the injury. Not only does this prevent medical problems from being overlooked, but it gives the worker the positive impression that his or her health is of prime importance.

### Return-to-Work Philosophy

After being evaluated for diagnosis and treatment potential, the worker is considered for possible functional activity. Being injured does not necessarily translate into being "sick." Therefore, options are discussed at the very first visit and return to work is implemented if possible. If immediate return to work is not possible, closely scheduled follow-up visits and a return-to-work plan is implemented.

### Quality Care

A worker who is sent to a highly reputable physician, or clinic or hospital, feels that no expense is being spared on his or her behalf. The worker has the attention of medical professionals and the injury receives full treatment potential. Even if the highest quality care is the most expensive, the avoidance of prolonged problems is well worth the initial investment.

Quality care also is monitored by measures of effectiveness and outcome. A practitioner who provides quality care can be identified by record keeping and outcome statistics for previous work injury experiences.

### Team Approach

The medical team approach involves many professionals and almost always includes a physician. In work injury, the team also includes representatives of the employer. The format of the team is one of medicine and industry working together. Communication, return to work, and proper reimbursement are maximized when all concerned agree on philosophy, process, and goals.

### Positive Approach

A visit to a medical practitioner can be a positive or negative experience. Even though medical professionals have a tight schedule and much responsibility, individual contact and positive attitude are every bit as important as competent medical techniques. The injured worker who perceives that treatment is being delivered with caring, competency, and interest in the results will be positively affected by the medical incident.

Assurances that general health prevails should accompany attention given to an injury and limitation of a particular work function. For example, an injured engineer released to work with a lifting limit of 25 lb will be pleased to know that his injury will not preclude him from being active in walking, mechanical controlling, and other activities of daily living and work.

## EXAMPLES OF DELIVERY SYSTEMS

Several injury care delivery systems have been developed. Small communities may mobilize existing medical forces to work as a team. Large cities with multispecialty clinics and hospitals may be able to offer more comprehensive services. Common components in these systems are a person who acts as coordinator, medical professionals who are familiar with the actual industry, and a coordinated system of evaluation, treatment, and reporting.

### Industrial On-Site Medical Teams

Many large companies hire physicians, occupational health nurses, physical therapists, and medically oriented safety officers. This facilitates positive results in three ways.

First, any injury can be treated immediately with emphasis on the needs of the employee. If the injury is beyond the scope of on-site medical professionals, there is an established referral base in the community for specialty treatment and evaluation.

Second, the confidence of having medical professionals on site increases the worker's ability to return to work safely. The worker knows medical care is available if the return to work produces increased symptoms. Also treatment and further evaluation are facilitated as time is not lost by the worker traveling to and from medical clinics.

Third, the actual site and method of injury can be inspected. This is important for ergonomic or safety intervention, which can prevent future injuries either to the individual worker or to the total work force.

**Occupational Medicine Physician**

While a physician can be known as an occupational medicine physician by various routes, there are two major categories currently involved in this specialty practice.

*Specialist*

Independence is one of the benefits of being a specialist not employed by industry or a large medical system. Most occupational medicine specialists wish to retain a neutral image. This allows the employer in industry to use a medical specialist who understands the needs of that industry and also allows the worker to know he or she has access to quality medical care oriented to his or her own successful health outcome.

Occupational medicine physicians often are designated for follow-up visits if evaluation in an emergency department has been required. If emergency treatment has not been necessary, often this physician is the first to see the injured worker.

This specialist physician has visited industry and knows the special problems and individual needs for work safety. The independent physician also will refer the worker to specialists for individual problems beyond his or her capacity.

*Designated Occupational Medicine Physician*

In some areas a trained occupational medicine specialist is not available. In this instance, a company may choose a general practitioner, family practice physician, internist, or other general care physician to act as the occupational medicine coordinator. The designated physician will see a mix of work injuries. Referrals will be made for more complicated problems.

In the event of emergency department visits, this physician often accompanies the patient and takes responsibility for follow-up for the work-injured patient. A designated occupational medicine physician can be neutral, highly professional, and effective in coordination of medical care.

## Occupational Medicine Health Systems

A health system indicates a multidisciplinary group with a variety of practice aspects. It is differentiated from small specialty groups by its ability to provide several methods of care.

### Clinic

Large, multispecialty clinics have the potential to meet occupational medicine needs. Because of the diversity of physicians and the numbers of specialities that may be involved, a medical coordinator is a must.

The coordinator may be a physician or an occupational health nurse who tracks all injured workers through the systems. This person is able to cut through medical system barriers; this may not be possible for anyone other than a representative of the clinic.

The coordinator is the professional that is familiar with industry and the needs of industry. This information is then communicated to individual physicians to assist them in making medical judgments regarding return to work. The occupational medicine coordinator ensures that systematic reporting and thorough communication take place.

One of the positives of the multispecial clinic system is that outside referrals are usually not required. A multispecialty clinic can effectively treat most problems within its own system. This facilitates the team approach.

### Hospital

Because many work-injured patients are treated initially in an emergency department, interested facilities are designated as occupational medicine specialty hospitals. Emergency department physicians are educated on the needs of industry and return-to-work philosophies. They have the ability to handle crises and refer appropriately if further testing is needed. They coordinate a team approach for initial diagnosis and treatment.

In occupational medicine hospitals, other departments, such as laboratory, X-ray, and physical medicine, assist the emergency department physician. An internal coordinator can be used to facilitate handling of all work-injured clients. There is most often an additional referral relationship with a comprehensive physician group. Overall care is maintained in the hospital system.

## "PERKS" FOR WORK-INJURED PATIENTS

Especially in the case of large medical facilities, the work-injured employee could be delayed or misdirected if certain precautions are not taken. Some of the measures instituted include the following:

- allotted parking space for injured workers
- special identification cards, which designate the industry, the name of the worker, and the type of procedure to facilitate reports, billing, communication, and return-to-work recommendations
- a system that reduces the waiting time to a minimum and assures immediate attention by an occupational health nurse or a physician
- one central report, which combines the evaluation and treatment recommendations of all professionals
- an immediate report, which is a compilation of the central report for immediate feedback to the industry, followed later by written communication
- simple referral procedures from one department to the next within the facility
- an employee or industry representative who is available for consultation or who accompanies the work-injured patient through the medical system

## ONE SYSTEM EXPLORED

Progressive concepts in occupational medicine have been developed by Peter Person, MD, and Jean Brisson, RN, for the Duluth Clinic, Duluth, Minnesota. One hundred and thirty-five medical specialists participate in a system that has allowed occupational medicine patients to be seen quickly, effectively, and comprehensively. Some of the aspects that have aided in this development include:

- universal forms for worker injury (Exhibit 11-1)
- universal forms for follow-up (Exhibit 11-2)
- universal preemployment screening examination forms
- coordinated physician care
- computer tracking to follow the worker through the evaluation and treatment process

**Exhibit 11-1** Worker Illness Physician's Report

WORKER ILLNESS<br>Physician's Report

**⊁⊰The Duluth Clinic, Ltd.**

Medical Record No. _____ Today's Date _____
Name _____ Date of Injury _____
Address _____ Date of First Visit _____
_____ Is further medical care necessary? Yes _____ No _____
Birthdate _____ Social Sec. # _____ Next Appointment Date _____
Occupation _____
Employer _____ Address _____ Phone _____
Insurance Carrier _____
How did injury occur? _____
_____
_____

Today's Diagnosis _____
Treatment/Medication _____
_____
_____

Was the injury or disease caused, aggravated, or accelerated by the patient's alleged employment activity? Yes _____ No _____
Did the patient have a pre-existing condition which affects the current disability? Yes _____ No _____
If yes, describe _____
Is patient likely to miss at least 60 work days? Yes _____ No _____ Can't evaluate at this time _____
Is permanent disability likely? Yes _____ No _____ Can't evaluate at this time _____
WORK STATUS:
Totally disabled from _____ to _____
Partially disabled from _____ to _____
Return to work:  M  T  W  T  F  S  S   Date _____  [ ] With no limitations  [ ] With work limitations
  Duration of limitations _____ to _____ Restricted lifting: _____ lbs.
  Restricted use of hand:  Right _____  Left _____  No use _____ .
  Wound care:  Regular job _____  Dry, clean job _____
  Other restrictions: _____
(If work within these restrictions is not available the patient is totally disabled.)

Physician Comments: _____
_____
_____
_____
_____
_____
_____
_____
_____
_____
_____

_____ Physician Signature _____

I authorize ⊁⊰The Duluth Clinic, Ltd. to release information on my injury or illness to my employer and employer's insurance carrier.

Signature: Name: _____ Date: _____

EMPLOYER COPY

*Source:* Reprinted with permission of The Duluth Clinic, Ltd, Division of Business and Occupational Health, Duluth, Minnesota.

122   WORK INJURY

**Exhibit 11–2** Worker Illness/Injury Follow-up

## WORKER ILLNESS/INJURY FOLLOW-UP

### ✂. The Duluth Clinic, Ltd.

Medical Record No. _____    Today's Date _____

Name _____    Date of Injury _____

Occupation _____    Date of First Visit _____

Employer _____

Diagnosis _____

☐   This is to certify that the above-named individual is under my care and will not be able to return to work until _____

☐   This is to certify that the above-named individual is under my care and is now released to return to work without restrictions as of _____

☐   This is to certify that the above-named individual is under my care and able to return to work as of _____
with the following restrictions _____
_____
_____

Above-named individual is scheduled for re-check appointment on _____

Comments/Instructions: _____
_____
_____
_____
_____

_____
Physician's Signature

White Copy: Employer     Pink Copy: Chart     Yellow Copy: OH Office

*Source:* Reprinted with permission of The Duluth Clinic, Ltd, Division of Business and Occupational Health, Duluth, Minnesota.

One aspect that has worked well for all physicians involved is the appointment of Dr. Person as the medical director. He has responsibility to act as liaison and coordinator for the comprehensive medical evaluations available. Meetings are held quarterly with the medical staff to review strategy, to facilitate coordination, and to review the progress of the occupational medicine department. In addition, discussions are held to promote communication with industry.

To aid in occupational medicine awareness, OSHA regulations are followed closely for changes in exposure limits, screening programs, and so forth.

The medical director and occupational health nurse are available to interact with industry to review any changes in work methods or aid in replacement of worker after effective treatment.

The nurse coordinator directs the activities of the occupational health nurses and communicates with industrial safety officers. There is frequent communication with medical personnel and industry to promote individual follow-up of injuries and medical care. Satellite clinics are supervised by the coordinator and education of occupational health nurses in these sites is maintained.

Additional occupational medicine goals are the following:

- To work with industry to determine its overall safety, health, and wellness needs.
- To review injury and illness records so that high-risk areas can be targeted and preventive goals can be established and met.
- To evaluate adjusted-duty jobs so that return to work can be facilitated with meaningful, modified duty at the worksite.
- To provide testing, consultation, and follow-up on mandatory employee screening.
- To assist the company in establishing policy and protocol for all its health and safety needs.

**Case Example**

Mr. Worker Brown reported sudden low back pain after fixing a fast-moving pulley. He immediately reported this pain to the supervisor who called the nearest satellite clinic to the industry. The coordinator asked that Mr. Brown be brought immediately to the clinic for examination. Following established protocol, the supervisor arranged for a foreman to take responsibility for operations in his absence. The supervisor then drove Mr. Brown to the clinic for evaluation by the medical staff.

Mr. Brown was first seen by the nurse coordinator who took a brief history of the injury, recording the type of accident, where it took place, and the type of symptoms that resulted. He was then seen by a physician who stated that this was a low back muscle strain caused by overexertion in the back flexion position. Mr. Brown was symptomatic when in the back flexion position or when doing active flexion-extension movements. However, he was able to sit comfortably in an upright position and walk short distances.

Mr. Brown was treated in the physical therapy department with cold applications and mild exercise. He stated he felt less symptomatic and able to return to work for the rest of his shift. The medical coordinator, knowing the extent of the injury, made arrangements with the foreman that allowed Mr. Brown to limit work activity to quality checks on the products. Rest breaks, as needed, were approved for the next three days.

Mr. Brown was able to return to work on each of the subsequent three days and went for a follow-up visit with the physician and occupational health nurse at the clinic after this period. There were no residual deficits from the muscle strain. He was then evaluated for full motion and ability and returned to full duty.

The important aspects in the case are that the clinic facilitated immediate care, that the industry supervisor was able to accompany the employee to the medical clinic and receive full information, and that the return to work was facilitated with both industry and medical approval.

In addition, the paper work was completed on occupational medicine forms, which were tracked through the nurse, the physician, and the physical therapist. The forms contained statements regarding return-to-status and were completed at the same time the notes were being written. Forms were in triplicate so that one copy could remain in the chart, one copy could go with the worker back to work, and one copy could be retained for insurance company purposes.

## PREROGATIVE OF INDUSTRY IN MEDICAL CARE SELECTION

With the advent of occupational medicine programs, directed services for industrial injury are available in many locations. Whether or not one of these programs exists in a community, industry is now taking a pro-active role in the determination of type of services and the actual service provider.

Tom Broderick, Safety Security Manager for Rust International Corporation, a major construction company, is an industry specialist. In an interview setting, Broderick stated that industries seek medical providers who provide high quality care, who are progressive, and who have a strong return-to-work emphasis. He is responsible for choosing medical providers in settings where Rust Engineering is managing a large construction project. He expects medical providers to be aware of industry's point of view.

Prevention of work injuries is the first goal of a state-of-the-art comprehensive loss control program. Workers are educated in safety concepts before beginning their work activities. Management provides constant reinforcement of safety practices. This preventative approach has greatly reduced the number of fatal accidents that were once accepted as a part of heavy industrial construction. In addition, many of the more severe types of accidents are being prevented.

When deciding on medical providers for a particular project, it is acknowledged that a number of employees will receive injuries. The programs for care after injury, therefore, are designed purposely to maximize early treatment of injury, to minimize long-term disability, and to promote a safe return-to-work approach.

Also promoted is the team concept with the employer and employee considered as much a part of the team as the medical providers. Safety managers need to be constantly aware of the status of any particular worker and his or her injury. It is this comprehensive communication that facilitates progress updates, reinforcement of proper care, and safe return to work. The communication flow includes written reports, telephone conversations, and meetings.

The following attributes of a system are preferred by Broderick:

1. A standardized coordination approach to management of the work injury.
2. A written plan of care with goals and timelines for aspects of healing, treatment, and return to work.
3. Constant computerized work tracking of the medical progress of the worker, which is retrievable by a worksite computer.
4. Regularly scheduled rounds involving medical care providers, industrial representatives, and, at times, the worker.

## SUMMARY

Quality medical programs that provide immediate care for the injured worker have improved the work injury management system. Not only have these programs met facility goals of increasing medical care volume, but also they have initiated contacts with industry that have provided beneficial services to the injured worker. Injured workers in this type of system are much more likely to have safe and early return to work without residual disability. Workers' compensation costs are thereby reduced, and the industry thrives with a healthy work force.

Immediate care takes full advantage of the medical profession's diagnostic and treatment skills. Rehabilitation programs are implemented far earlier with good results. The return to work is no longer viewed by medical professionals as a return to an injurious environment. The future is positive for further interaction of medicine and industry in early care of the injured worker.

# Section 12

# Role of the Physical Therapist in Early Intervention

*Susan J. Isernhagen*

## INTRODUCTION

The goals for care of acute musculoskeletal injury are:

- To promote healing and comfort with treatment modalities.
- To incorporate initial exercise to assist in the rehabilitation process.
- To begin education that provides a client with methods and information necessary to take an active role in the restorative process.

Prevention of future reinjury is an important aspect of the total process. The earlier intervention after injury is begun, the more successful the treatment and safe return to work will be.

General categories of injuries seen by the physical therapist for immediate care are sprains, strains, and severe symptoms of cumulative trauma. Back injuries are the largest single diagnostic group, and they are complex in their presenting symptoms and causes. Following low back injury, injuries to the neck, shoulder, hand and wrist, hip, knee, and ankle are most commonly seen.

There is a general consensus in occupational medicine, and particularly in physical therapy, that although a great need exists for immediate musculoskeletal care, this is an underdeveloped area of the entire medical plan. If injuries do not receive proper attention initially, healing is slowed and secondary problems can occur. The following are counterproductive to early and safe return to work:

- inadequate initial treatment
- prolonged bed rest without full medical necessity

- exclusive treatment with drugs rather than incorporation of physical means
- lack of attention to an underlying medical condition that can cause reinjury
- lack of activity, which causes deconditioning and proportionately greater lost time from work

Several physical therapists involved with early intervention were surveyed regarding their experience in facilitating healing, prevention of reinjury, and early return to work of the injured employee. Their opinions are incorporated into this section.*

## DELIVERY SYSTEMS FOR IMMEDIATE CARE OF INDUSTRIAL INJURIES

### On-Site Physical Therapy

The closest working relationship in immediate care integrates the physical therapist within the facility. Only the largest of industries currently have physical therapists on staff. Often these physical therapists have many additional duties, such as designing and implementing health and fitness programs, providing education to employees, and performing ergonomics analysis.

If a physical therapist is present on site, there is immediate access. Injured workers can be evaluated within minutes of the injury and the treatment modalities applied if indicated. The setup of the industrial physical therapy department is often basic, with equipment similar to that used in athletic departments to treat sports injuries. Usual treatment modalities include cold packs, electrical stimulation, and ultrasound. In the case of special or severe physical injuries, referrals are made to specialized physical therapy providers.

The physical therapist within industry also is in an excellent position to monitor a return to work after treatment. The employee can be checked several hours after injury and also on a daily basis.

---

*Participants in the survey were John Brickley, PT, Polinsky Medical Rehabilitation Center, Duluth, Minn; Janice Culliton, PT, St Mary's Medical Center, Duluth, Minn; Richard Ekstrom, MS, PT, St Luke's Pro-Health Occupational Medicine Department, Duluth, Minn; Allen Holm, PT, McKennan Hospital, Sioux Falls, S.Dak.; Duane Saunders, MS, PT, Park Nicollet Occupational Medicine Department, Minneapolis, Minn; Andrew Wood, MS, PT, General Mills, Inc, Minneapolis, Minn.

## Proximal Physical Therapy

When a physical therapist has an office in close physical proximity to industry, early intervention also is possible. The physical therapist would see a patient, apply the correct treatment, and be able to return the injured worker to work with almost the same expediency as the physical therapist within the industry. This satellite physical therapy service provides care on a timely basis with little delay for initial service. For recurring treatments, additional appointments would be scheduled to avoid waiting time. Staffing levels of these clinics would be adjusted according to industry needs.

## Contract Services

Many hospitals and medical clinics have established occupational health services that provide immediate evaluation, treatment, and follow-up. These occupational health services, although not located on site, usually are effective in providing "one-stop shopping" so that all the medical needs of the injured worker are met. In addition to physical therapy, the services of the X-ray department, laboratory, and pharmacy are available.

In addition to the comprehensive occupational health contract, physical therapy clinics can provide "one-stop" rehabilitative care for the injured worker. A continuum of immediate treatment, work hardening, and functional evaluation can be provided by one facility. If all workers with musculoskeletal injury are seen, total patient volume may rise, but lost-time days usually decrease.

McKennan Hospital in Sioux Falls, SD, has taken the concept of immediate care of injury several steps further, interfacing early intervention with appropriate education and rehabilitation. There is a continuum of programs until safe return to work has been achieved. It is modeled after the well-accepted three phases of the cardiac rehabilitation process.

For back care rehabilitation, in phase I injured workers are seen for emergency department evaluation and referral to physical therapy for acute care. Phase II is education. If the injury continues to interfere with work, phase III, back fitness rehabilitation, is implemented. If problems continue, phase IV is work tolerance assessment and conditioning. A worker may be discharged at any point successful return to work has taken place. This phased program succeeds because of immediate referral and continued attention to the worker until results are positive.

## TRENDS IN IMMEDIATE CARE

There is a growing awareness in all levels of medical care that work injuries are best seen by specialists trained to deal with occupational injuries. These occupational medicine specialists have a knowledge of the specific injury and can facilitate early and safe return to work. A "sports medicine" philosophy is adopted. Occupational injuries receive early evaluation including the use of modalities that effectively reduce symptoms and facilitate return to function as soon as possible.

The physical therapists involved in early intervention estimate that only 5% of musculoskeletal industrial injuries are seen in the immediate care situation. Although it is accepted in industry, and agreed on by physical therapists and physicians, that immediate care is very helpful when instituted, the current availability of services falls short of meeting that demonstrated need.

## TYPES OF PHYSICAL THERAPY INTERVENTIONS IN IMMEDIATE CARE

### Modalities

The most useful modalities are those that reduce pain and swelling and promote healing. Among the most highly used are cold, high voltage galvanic stimulation, and ultrasound combined with proper resting of the part. If an extremity is injured, other principles such as compression and elevation are used. After the acute stage, manual therapy, traction, and hydrotherapy with exercise are given for both symptom relief and restoration of function.

### Selective Rest or Immobilization

To protect the part from early reinjury, rest of the injured part can have a curative effect when prescribed properly. Rest is helpful for muscle spasm or instability.

Immobilization can aid in assuring rest by physically restricting motion. Involuntary muscle contraction, which can be postural or stabilizing in nature, also will be reduced if supportive immobilization is used. Types of immobilization most commonly used in these settings are neck or low back braces or neck splints to reduce motion; wrist and hand splints, which may allow motion of uninjured joints but restrict motion in the injured joint; and bracing for weight-bearing joints such as knees or ankles.

Immobilization should be evaluated on a daily basis so that rest ends when healing has been accomplished. Functional activities, such as return to work or activities of daily living, should be carried out even though one part of the body may be resting or immobilized. This prevents deconditioning and avoids the negative "sick" image that a worker may develop.

### Exercise

If the injured part is immobilized, isometric exercise may be instituted to improve muscle tone of and blood flow to the injured area. If immobilization or rest is not required, exercise instruction can be graded to allow relief of pain and improved mobility and to prevent deconditioning. If there isn't a return to full function within a few days or weeks, exercise may need to be progressive in nature, rehabilitative in character, and result in a work hardening type program.

### Modification of Worker Methods or Work Practices

Educating workers in the treatment of their specific physical injury is of utmost importance. This facilitates the healing process and the return to function. In addition, the worker learns how to return to work without aggravating the part. For example, a worker with a back injury may be able to return to work if bending or lifting restrictions are followed.

Another form of education is the review of the accident or injury. Unsafe body mechanics, unsafe work practices, or improper match of work to worker may be identified. Education is very well received when the employee is recovering from an injury. He or she no longer feels like "that will happen to someone else, not me." New or safer methods of work and worksite modifications are more easily implemented at this time because the impact of poor work practices is visible.

## RETURN TO WORK

The therapists providing immediate intervention estimate that 50% of injured workers can be returned to work on the same day they are first seen.

Treatment can continue, if indicated; but uninterrupted employment also is possible.

Even if immediate return to work is not possible, early intervention assures return to work is accomplished on a more rapid work progression than that seen when bed rest is prescribed for the injured worker. Good communication among the physical therapist, physician, and employer facilitates progress. In most cases, return to work can usually be accomplished within two to seven days. When lost-time days are minimized, the ultimate outcome is usually positive.

To prevent return to work from being instituted too rapidly, discharge criteria are important. Undertreatment of a serious condition is as important to avoid as overtreatment.

The goals of treatment should be the return to safe work function. If an underlying, predisposing physical problem is discovered, it also should be addressed. Reaching a safe functional level, however, does not mean a restoration to physical perfection. Unrealistically high expectations of treatment outcomes are often the reason behind overtreatment.

## TRENDS AND FUTURE CHALLENGES

The availability of immediate care for the injured worker is increasing. Professionals involved include physicians, occupational health nurses, physical therapists, and others involved in immediate treatment. The increase in early treatment is seen both as a cause and an effect of the improved relationship between medical personnel and industry. To facilitate the early return to work, which is part of immediate treatment, there is a need for improved understanding of medical problems by industry professionals, and of the work by medical professionals.

There is increasing awareness of the role of the physical therapist in the injury recovery process. A good relationship between the physician and the physical therapist will allow not only early referral, but also better communication and agreement on the return to work. Physical therapists who work directly with injury will continue to communicate with physicians designated by the industry. Although the percentage of workers who have access to immediate care is currently low, it is expected to increase in the future.

Future challenges to professionals involved in immediate musculoskeletal care include meeting the following needs:

- better communication among all occupational medicine providers involved in immediate care

# Role of the Physical Therapist in Early Intervention 133

- improved location of physical therapy departments, either on site or in closer proximity to industry
- increased numbers of physical therapists trained in industrial physical therapy concepts
- identification of high-risk work or worksite, which may contribute to injuries
- research that compares the cost of early intervention to that associated with traditional medical treatment of injured workers
- research that compares the long-term physical and safety results of immediate treatment versus those of traditional medical treatment

# Section 13

# Stabilization Principle for Physical Management of Work Injuries

*Arthur H. White*

The vast majority of time, workers with back injuries return to work after relatively simple treatment measures. With just time alone 94% of back-injured people return to normal; 6% are at risk for the work injury to become a chronically disabling injury. Also, people with back injuries are at five times the risk for recurrent injury as those who have not had previous back injuries. Therefore, it is incumbent on those dealing with work-related low back injuries to institute preventive measures in as simple and economical fashion as possible. Back school can provide such prevention. Simple exercise programs can be instituted at home or at the worksite. Stronger measures are needed when major levels of debility exist. Employees who have gained considerable weight, have not exercised for years, and are returning to heavy labor jobs should be reconditioned to a level that provides a good margin of safety.

When time and simple education and exercise do not return a patient to normal, additional formal training is required. Work hardening and gym programs may be successful over a 1- or 2-month period. These types of programs should include stretching, strengthening, and education. Body mechanics for activities of daily living and work-related activities should be stressed and practiced.

Stabilization training is a specialized form of body mechanics, posture, and movement and exercise. This type of training is relatively complex, time consuming, and difficult for both the employee and the instructor.

The principle of stabilization training is a retraining of all the muscles available to control the motion and position of a diseased spinal segment. Theoretically, it is possible to place any one spinal segment in a pain-free, well-aligned, balanced position and hold it there for any particular work task. Coordinating the muscles involved is foreign to most people. For example, most people are unable to contract one gluteus muscle at a time. Most people, when asked to tighten their abdominal muscles, actually "suck in their abdomen." Training some people to push their abdominal muscles out, thus tightening the muscles and stabilizing the diaphragm, and to continue to breathe at the same time can be very difficult. The latter activity is called abdominal bracing. The combination of abdominal bracing, gluteal coordination, straight-back bending, and paraspinal muscular alignment and splinting forms the basic principles of stabilization.

How does a person learn to use all of these muscles in this highly coordinated fashion? It takes time and practice. A person must learn to use these muscle groups one at a time, to use them independently, and, ultimately, to coordinate them all together as increasingly difficult tasks and loads are applied.

Each person has a different background in this type of coordination, depending on his or her previous athletic endeavors, experience with dance, breathing training, and lifting experience. Most people will start this type of training lying supine. They practice tightening the abdominal musculature by doing partial sit-ups. They find their comfortable neutral, balanced position with the help of a trainer by doing pelvic tilt positions, altering the spinal curvature until the best theoretical and actual position is experienced. That position of neutral balance is then held with the muscles above described while the patient raises one arm and then one leg, and then both arms and both legs, still maintaining the neutral, balanced position. Loads can then be applied in the form of weights or a patient's balance can be offset with the use of Swiss gymnastic balls.

Once stabilization is accomplished supine, then the training is done prone, kneeling, and, ultimately, standing. It takes weeks of training for some people to advance even to a kneeling stage.

Once standing stabilization is accomplished, the person's work retraining can begin. Job simulation is done maintaining the stabilization principle. If the work requires no more than the person's being able to stand at a bench or sit, the retraining is simple. The worker and trainer can practice finding the neutral, balanced position standing using the conventional props of putting one foot up on a step stool or railing. Appropriate weights can be lifted from this position while the trainer ensures that all the stabilization muscles are doing their job. Sitting stabilization is equally simple, as the trainer ensures that the neutral, stabilized position is being maintained as the worker sits in various chairs and positions. Ergonomic changes can be made for sitting and working so that the stabilization muscles can be rested while appropriate external supports take over the job of maintaining the spine in its pain-free, neutral, balanced position.

With jobs that require heavy bending and lifting or strong bursts of energy or receiving blows, the training becomes much more technically difficult. Up to this point in training, balance and coordination have been stressed more than brute strength. To stabilize against the forces of heavy lifts or external blows, the stabilization muscles must be able to maintain spinal position with equal forces and sometimes for a long time. Accordingly, months of progressive strengthening of stabilization musculature may be required before such heavy training can be attempted.

Forces are slowly and progressively applied in the form of lifting or delivering blows (eg, catching a medicine ball) while stabilization is achieved. As with all stabilization training, the patient must remain pain free during the activity and not suffer a recurrence of symptoms for 24 hours after the training.

In this fashion, there is no end to the stabilization capabilities that a person can accomplish. This is typified by occupations of professional football, boxing, and competitive weight lifting. With injured workers, however, there is a disease process or damaged spine that has its limitations. A good clinician should be able to determine what levels of return to work can be expected through stabilization training. It is simply an equation relating the underlying disease process (spinal damage) with the job required (physical demand level) and the worker's ability to protect himself or herself (stabilization capacity). High levels of stabilization training will obviously return more severely injured workers to the same job. At some point, the spinal injury is too severe to allow a person with any level of stabilization to return to a given job. At that point, the job needs to be changed.

In summary, stabilization training is an additional valuable but technically difficult tool that can be used to increase the work capacity of any spinal-injured worker. Because of the time and complexity of training required, other more simple tools, such as time, education, body mechanics, strengthening, flexibility, and work hardening, should be considered first.

# Part V

# Functional Capacity Evaluation

Functional Capacity Evaluation (FCE) is the comprehensive, objective testing of a person's ability in work-related tasks. It is musculoskeletally oriented so that as performance is rated, both abilities and physical limitations can be noted. It is relatively new in the field of physical and occupational therapy, but it has made a large impact on the return-to-work process of injured workers.

The FCE is performed on a one-to-one basis for several hours of intense evaluation. The purpose of the test is to stress the person's physical abilities to the maximum for accurate documentation regarding work and activities of daily living. It should describe full function of the injured worker, as well as limitations, so that further reinjury can be prevented. In the process, injured workers are educated on their own abilities and limitations to facilitate a more pro-active role in their return-to-activity process.

Functional capacity evaluation is an integral part of the return-to-work process. It forms a medical basis for return-to-work conclusions that allow appropriate productivity, with modification of physically contraindicated work activities. It interfaces with information from many professionals on the return-to-work team. Positive reinforcement has come from professional areas, such as industry, insurance, workers' compensation, and medicine. It is a base component of work injury management systems. A comprehensive, accurate FCE can be put to optimum use. It will be clear and assistive to all who need the functional work information.

## Section 14  Functional Capacity Evaluation

Many of the parameters and philosophical decisions that need to be made when adopting or designing an FCE are indicated. Although there are many variations, there are certain basic qualifications that should be present in the evaluation and that the examiner should possess.

In addition, three addendums are provided to enhance the understanding of the use of functional evaluation.

### —1. Disability Determinations

The potential use of FCE in case resolution is discussed. Disability settlements, from the point of view of an expert vocational evaluator participating in disability hearings, are described.

### —2. Cost Savings in Four Cases

Cost savings are realized when FCE is used appropriately in return to work. Four case examples are presented.

### —3. Functional Capacities Assessment Research: The Relationship of Age and Gender to Functional Performance—Patients and Uninjured Subjects

Data from the Polinsky Medical Rehabilitation Center regarding age and gender variables and their effect on functional performance are reported.

# Section 14

## Functional Capacity Evaluation

*Susan J. Isernhagen*

## OVERVIEW

### Definition

**Functional:** meaningful, useful. In this context, functional indicates purposeful activity that is an actual work movement. Functional implies a definable movement with a beginning and an end, and a result that can be measured.

**Capacity:** maximal ability, capability. Capacity indicates existing abilities for activities including the maximal function able to be used.

**Evaluation:** systematic approach including observation, reasoning, and conclusion. Going beyond monitoring and recording, the evaluation process implies an outcome statement that is explanatory, as well as an objective measurement of the activity.

A functional capacity evaluation (FCE), when applied to physical measurement of work or daily activity, requires the following:

1. The evaluation should be done by a professional with the ability to monitor, record, assess, and design an outcome statement.
2. The evaluation should include productive activities that are taken to completion. Both the process and the ability to complete a task should be evaluated.
3. In evaluating the function, full effort is required.
4. Any FCE should be done within safe medical parameters.

## Return to Work

When the FCE shows that the physical abilities are present and match the job (per job description), then a return to work is facilitated. The FCE takes the question mark out of the ability and potential reinjury category.

## Referral to Work Hardening

An FCE is the ideal evaluation to indicate whether work hardening is necessary and appropriate. In addition to showing deficiencies of physical ability, a good FCE also will show the physical reasons for those deficiencies. This facilitates planning a work hardening program, as both the physical problems and the work-related tasks are known.

## Job Modifications

Physical conditions may be present that prevent certain tasks from being performed safely. Job modification should then be instituted to ensure that these unsafe activities are eliminated. Options would be to modify the jobsite (eg, raising work station) or to modify the job (eg, allowing modified positions or repetitions for a particular activity).

## Disability Determination

In cases of litigation regarding disability from an injury, an FCE can clearly delineate the functional capacity. This can be combined with the physician's estimate of disability.

# FUNCTIONAL CAPACITY EVALUATION PARAMETERS

## Relationship to Work

Assessment of return-to-work capabilities is the prime purpose of FCE. Therefore, the test must be designed to incorporate pertinent work activities and functional tasks representative of stresses at work. The match of work task design and actual work will decrease the "inferential leap" made by the examiner and also will demonstrate the relevancy of the test to the injured worker.

**Example:** Joe has recovered from a severe right shoulder injury. He is stabilized and has reached maximum medical improvement. His job requires heavy work with the right shoulder in lifting, carrying, and working on equipment in an overhead position.

The traditional evaluation provides accurate information but no relevance to work. One might suppose Joe would be unable to do heavy work because of the shoulder tightness and strength deficits on the right side compared to the left side.

With the development of FCE, important questions can now be addressed: Can Joe go back to work? When? In what capacity? Functional testing shows that in a work relationship, lifting and carrying are adequate for the job. The only limitation is in prolonged reaching for equipment controls in an overhead position. If stress occurs, this position can be modified by using a stool to reach the equipment. Return to work was facilitated.

Table 14-1 compares the traditional approach with the FCE approach.

## Professional Evaluator

An accurate FCE requires professional skills for performance of the evaluation. A look at differences in gait evaluation illustrates this point. An unskilled person could accurately document that a person is walking with a limp. It would even be possible for this person to determine which side appeared weak or painful and further classify the limp as mild, moderate, or severe. The physical therapist, on the other hand, could identify a gluteus

**Table 14-1** Comparison of Traditional and Functional Evaluations*

| Evaluation—Traditional Approach | Evaluation—FCE Approach |
| --- | --- |
| Right shoulder:<br>Range of motion decreased by<br>• 30 degrees in flexion<br>• 30 degrees in abduction<br>• 20 degrees in external rotation<br>No limitation in internal rotation, abduction, or extension<br>Manual muscle test (MMT) indicates grade 4+ shoulder muscles<br>Cybex testing indicates right/left ratio represents right shoulder being 85% of left shoulder muscle testing | Can lift floor to waist 50 lb<br>Can lift horizontally 80 lb<br>Can lift waist to eye level 40 lb<br>Can carry two-handed 85 lb<br>Overhead position for unweighted work can be sustained for 1.5 minutes but can be tolerated infrequently because of stretch of shortened right shoulder structures |

*Condensed evaluations.

medius weakness with additional findings of tightness in hip external rotators. Weaknesses in the push-off stage of gait might be noted. Further professional examination would reveal exact characteristics of the lower extremity regarding tightness and weakness.

If a thorough, accurate FCE is to be performed, the therapist must have the necessary skills and, also, must use them to their fullest extent.

Competencies necessary for FCE include a thorough knowledge of the neuromusculoskeletal system. This includes muscle, nerve, soft tissue, and bone physiology, along with normal and dysfunctional characteristics of each. Kinesiology, pathokinesiology, and cardiopulmonary implications are necessary in the evaluation of dynamic movements. The physical therapist and orthopedically trained occupational therapist most often have this professional preparation. Specific training in FCE also is recommended.

**Therapist Assessment versus Worker Report**

If the injured worker could assess accurately his or her capabilities, there would be no need for FCE nor would the physician have any difficulty medically releasing an injured worker to go back to work—there would be justification in merely asking the worker. Experience has shown, however, that injured workers are not always able to specifically or accurately estimate their own functional ability. Therefore, the therapist must use physical evidence of capability and limitation in documentation.

**Setting**

Functional capacity evaluation, because it is a medical test, should be performed in a medical setting. Neither the workplace nor the home is neutral ground. In these settings there may be psychological stresses to perform well or poorly. Functional capacity evaluations should be done in a clinic setting, with specific equipment and control of external variables that might influence the test results.

Because standardization is an important aspect of the FCE, the equipment and its placement is important. Selection of the items to be tested and arrangement of all testing equipment should be done in advance of an assessment. A designated space in the department allows smooth transition from one test item to the next and emphasizes the professional aspect of the testing situation.

## Length

### Standard Evaluation

For a standard FCE, the recommended time is four to six hours. The most reliable format is two consecutive days, with the most critical items being repeated on the second day. The evaluation time may be longer on the first day than on the second, or it may be allotted equally between the two days. The two-day format allows for retesting for accuracy and evaluating the effect of the first day's work testing on the worker. In some cases, the evaluation on the second day will reveal that the worker has increased physical symptoms (muscle spasm, joint swelling) from work items done on the first day. In other cases, the worker may show increased abilities on the second day (because he or she has worked through the fears and caution) or may show performance on the second day consistent with that on the first day. Therefore, the second test day is highly important to evaluate the effects of the work so that recommendations for day-to-day work activity can be made.

There is a tendency on the part of some referral sources or evaluators to test only the functions directly related to the injured part; that is, only upper extremity functions for a shoulder injury, or only lifting and carrying for a back injury. Because a true FCE is a report of ability limitation in all possible work activities, a comprehensive, standardized format will give the best work potential information. This will prevent any secondary problems from going unrecognized and also allow a strong listing of abilities (which are the true return-to-work necessities) as well as limitations.

### Single or Specific Performance Rating

In select cases, only limited functional testing will be required. Functional test items can be used very effectively. However, the limited nature of the result should be emphasized; and limited testing should not be confused with holistic FCE.

> **Example:** A physician may be completely confident about an employee's full return to work. The only questionable aspect may be the maximum amount of weight that can be lifted safely. The referral would be made for a one-time evaluation of lifting. Return to work can then be accomplished rapidly.

Table 14-2 highlights limited functional testings.

This type of testing also is appropriate for pre-employment screening. It allows specific functional activities to be addressed economically and quickly.

**Table 14–2**  Limited Functional Testing

*Characteristics*
- The test is task-oriented rather than total-person function oriented.
- Strength is evaluated without endurance components.
- Format is for one-shot administration.
- Report is brief and specific.
- Cost is low (compared with FCE).

*Positive Aspects*
- There is low time involvement, as well as lower cost (compared with FCE).
- Unnecessary activities are eliminated.
- A brief and timely report is possible, which speeds case resolution.

*Potential Problems*
- A secondary physical problem may not be identified. Example: If a progressively degenerative knee joint had changed lifting body mechanics thereby causing a back injury, the one-time brief lifting test may overlook the knee dysfunction that would be discovered during full FCE.
- Medical effects of deconditioning may change the client's ability levels. Findings of weakness, tightness, or dysfunction at areas other than the injured part would not be discovered.

## *Work Capacity Evaluation or Work Tolerance Screening*

In chronic injury cases or in the situation of a very demanding job, a longer evaluation period and testing may be indicated. This is often referred to as work capacity evaluation. The definition here describes a 1- to 2-week testing period, 8 hours per day, 40 hours per week.

> **Example:** Because of long and difficult temporary disability for L4-5 disk rupture and subsequent surgery, a client may be deconditioned and fearful. A 1-week evaluation with simulation of actual work hours and job demands will identify capability for continued work activity.

Table 14-3 describes the work capacity evaluation. Despite the potential problems, work tolerance screening continues to be very valuable, especially for the long-term chronically injured worker.

## DESIGN OF THE FUNCTIONAL CAPACITY EVALUATION

The design of the FCE should incorporate all aspects of functional capacity and, in turn, should accommodate their respective evaluation.

**Table 14–3** Work Capacity Evaluation

*Characteristics*
- The test is oriented toward evaluation of total body function, but usually has a vocational, as well as a medical, component.
- Endurance and alternative methods of work are emphasized.
- The report often is long and involved and may cover psychological as well as physical behaviors.
- Cost is higher (compared with FCE); but, because of length and lack of evaluator-client intensity, the test is less costly per hour (compared with FCE).
- Test is often monitored by paraprofessionals, although overall supervision is by professionals.

*Positive Aspects*
- Ability to perform work tasks at a daily and weekly endurance level can be shown.
- The act of returning to a work format will increase the worker's abilities and begin to restore lost confidence and endurance.
- The work periods are similar to actual work, so compliance and attitude also can be evaluated.

*Potential Problems*
- The client may already have the ability to return to work. A less costly and more time-efficient FCE could effect an immediate return to work.
- Prevalent use of work capacity testing when not indicated can lead to the impression that high time and cost commitments are necessary, thus delaying referral until only the most troublesome, chronic clients are seen.

---

Table 14-4 shows the essential assessment categories, and each is described in the sections that follow.

## Strength

Important information for employers and counselors is the injured worker's functional strength. How many pounds can the worker lift? How many pounds can the worker pull? Can the worker lift a 100-lb box off the shelf and load it onto a truck?

Workers without the basic strengths required for their jobs are recognized as having an injury potential. Without basic strength in a functional activity, stress will be transmitted to accessory muscles and the skeletal structure. Overuse or strain of muscle is possible, as well as further damage to the musculoskeletal system.

**Table 14–4** Essential Assessment Categories in the Functional Capacity Evaluation

Strength (dynamic)
Endurance
- Muscular
- Joint
- Cardiopulmonary

Pace
Coordination
Balance
Safety

Although isometric testing can produce a value in pounds, and isokinetic testing can provide torque values, they are not directly correlated with realistic, dynamic lifting. For the purpose of functional lift capacity, actual lifting should be done by the worker to a maximum, safe level.

Strength testing should be done for a sufficient number of repetitions to ensure that primary muscles continue to be the prime movers. If these muscles weaken after one or two repetitions and accessory muscles are used, or if safe body mechanics are not applied, this would indicate that the strength to maintain an activity is not present.

In addition to lifting, the most desirable strength capacities are carry, push, pull, and hand grip force.

**Endurance**

Endurance in FCE testing is the capacity to continue an activity. This is very helpful in correlating repetitious activity at work with the functional testing. The higher the percentage of strength used in an activity, the less the endurance. Although ratios are being studied for normals, the ratio of number of repetitions to a percentage of maximum strength does not necessarily apply to injured workers. Injured workers have physical problems, and these physical problems will make them less able to be categorized.

Muscle physiology research indicates that some people have greater capacity for strength tasks and some have greater capacity for endurance tasks. Just as joggers seek different activity from sprinters, heavy load lifters may have different muscle characteristics than assembly line workers. Matching strength and endurance of the workers for their work can aid in prevention of injury.

Another factor that will influence endurance relates to the abilities of the joints to tolerate movement and compression. If there is joint dysfunction because of trauma or arthritic changes, ability to perform repetitions will be diminished. This type of limitation will be important to note, as further damage could be caused if restrictions are not followed.

Cardiovascular and aerobic capacity also will affect endurance, particularly heavy endurance tasks such as lifting, stair climbing, and manual materials handling.

Other factors that may affect endurance are sensation, swelling, and circulation.

## Pace

Pace is often predetermined by the job. An assembly line worker must be able to maintain a speed equal to the speed of the line. In addition, workers who do piece work and are paid by the items they produce may work at a fast pace to increase adequate compensation for their work. Also, people have a natural pace that they set internally. Walking pace, working pace, and even talking pace differ significantly among people. Therefore, the person's normal pace is important to establish during functional testing. For specificity, the work pace can be added to the FCE testing situation for comparison to the self-set pace.

## Coordination and Balance

Testing hand coordination is important in the FCE for two reasons. First, upper extremity coordination is required in most work and activities of daily living. If impairment is found, the degree of the problem should be documented and methods of making changes should be recommended. Secondly, the documentation that ability exists is extremely helpful in cases when an employee who formerly did heavy labor cannot return to that work. The ability to have transferable or additional skills is crucial. By documenting abilities, a counselor or employer knows what the employee can do and a new job can be created or found.

For many clients, hand coordination and repetitive motion make up a large part of their job. Hand and upper extremity testing should include testing for coordination, strength, sensation, position, and repetition. The testing also may incorporate need for splints or assistance devices.

If the sitting position is used for performing the hand coordination tests, sitting tolerance and working habits can be observed. This is particularly necessary in patients with lower back or upper back and neck dysfunctions.

Similar to pace, if work is required at a coordination level above the normal ability, tension and stress can result and injury can occur.

Balance is a skill involving the lower extremities, back, neck, cerebellar, and vestibular function. In musculoskeletal injuries (the predominant injury seen in the injured worker) joint proprioception or muscle weakness problems can be evident in balance activities. If a person is having difficulty with balance, he or she is more likely to be injured, and his or her strength and endurance will be of little value.

**Safety in Work Motions and Body Mechanics**

Medical professionals are aware of unsafe procedures that may cause injury. In the treatment phase, the injured worker's safety is ensured. Also, he or she is helped to understand how to maintain his or her own safety. Similarly, this ethical attitude is important when doing functional capacity testing.

If the injured worker uses safe body mechanics, this should be noted. Testing can continue until maximal function is reached. If unsafe movements are used during testing, then correction should be attempted. If the person is unable to maintain a safe, correct method, then the activity should be stopped and the reason documented. The documentation would include the test results and need for further intervention.

If the client lifts unsafely and is injured, or claims to be injured, the therapist would be ethically questioned regarding knowingly allowing a client to perform an unsafe activity. Legal liability for that injury could be placed on the therapist.

In any event, the method of lifting techniques should be described when the report is sent. In addition, recommendations should always be made if unsafe body mechanics need to be addressed in a work hardening or body mechanics training session.

The insistence on safety during testing will be well appreciated by the client, the employer, the insurance company, and the physician. The establishment of safe procedures gives credence to professional conduct.

## COMPONENTS OF FUNCTIONAL CAPACITY EVALUATION

The components of the FCE are the most commonly required tasks specified on workers' compensation medical forms and also the most prevalent activities at work and in daily living. When combined they will give an overall picture of the person, complete with abilities and limitations. All muscle groups and joints are stressed several times in a variety of positions and activities. The total result is an individual score for each item, but a global physical evaluation of the patient also can be made.

A generic standardized FCE comprises (1) the collection of patient data (Form A; Exhibit 14-1); (2) the FCE report (Form B; Exhibit 14-2); (3) the FCE test (Table 14-5); and (4) the summary report (Form C; Exhibit 14-3). In an actual case study information derived from the evaluation is condensed and recorded on Forms A and C; specific test results are recorded on Form B. As examples of FCE test items, Figure 14-1 shows a test for hand coordination; Figure 14-2, a test for dynamic pushing ability.

## EVALUATOR PERSPECTIVE

Philosophical issues can arise during intense evaluation interaction with injured workers. Thinking through some of the attitudes that might need to be assumed is helpful. The following sections describe often-discussed approaches and preferred philosophies.

### Medical Problems Do Not Equal Functional Loss

Medical professionals are geared to finding the "medical problem." Logic indicates that identification of the problem is the first step in the goal of "healing." Once the patient is healed, however, the patient's physical problem should be deemphasized to rehabilitate that patient to a functional level. Even though this change of philosophies has taken place with rehabilitation for some patients (ie, those who have suffered a stroke, spinal cord or head injury, or neurological condition), more difficulty has been experienced in applying these philosophies to patients who have suffered musculoskeletal injuries.

In FCE the ability, not the disability, should be emphasized. The return to work is only possible when an employee demonstrates ability to perform work activities. Furthermore, it should not be presumed that a specific physical problem is necessarily going to be a functional problem as well. If musculoskeletal normalcy or perfection were required for work and productive living, there would be a very limited place in life for most people. This would be especially true for people with arthritis, scoliosis, or amputated limbs, or for any person who shows physical aspects of aging.

**Exhibit 14–1** Functional Capacity Evaluation: Patient Data

## FUNCTIONAL CAPACITY EVALUATION: PATIENT DATA

FORM A

*History*

Date:
Name:
Address:
Telephone No.:
Referral Source:
Physician:
Diagnosis:

Date of Birth:
Date of Injury:
Type of Injury:
Total Time Off Work:
Occupation:
Employer:
Insurance Company:

Pertinent Surgery:
Previous Treatment:
Current Medications:
Patient Report:
 • Functional Level:
 • Pain Level:
Work Status:

*Physical Assessment Summary*

Strength and Motion:   Comments:
(all motions within
normal limits unless
specifically noted here)

Neck and Back:

Upper Extremities:

Lower Extremities:

Blood Pressure:
Resting Pulse Rate:
Respiration Rate:
Gait:
Posture:
Coordination:
Balance:
Movement Characteristics:
(speed, smoothness, posturing)

*Source:* Copyright © 1988, Susan J. Isernhagen

Exhibit 14-2  Functional Capacity Evaluation Report

## FUNCTIONAL CAPACITY EVALUATION

FORM B

| Item | Frequency ||||  Restrictions | Comments |
|---|---|---|---|---|---|---|
| | 0% - 5% | 6% - 25% | 26% - 50% | 51% - 75% | 76% - 100% | | |
| Weight Capacity in Pounds | | | | | | | |
| Lift: Floor to CG | | | | | | | |
| CG to eye level | | | | | | | |
| Horizontal | | | | | | | |
| Push: Dynamic | | | | | | | |
| Pull: Dynamic | | | | | | | |
| Carry:         R | | | | | | | |
|                    L | | | | | | | |
| Front—at CG | | | | | | | |
| Handgrip:  R | | | | | | | |
|                    L | | | | | | | |
| *Flexibility/Positional* Elevated Work | | | | | | | |
| *Trunk Flexion Work* Sitting | | | | | | | |
| Standing | | | | | | | |
| *Unweighted Rotation* Sitting | | | | | | | |
| Standing | | | | | | | |
| Crawl | | | | | | | |

**Exhibit 14–2 Continued**

| | | | | | | | | |
|---|---|---|---|---|---|---|---|---|
| Kneel | | | | | | | | |
| Crouch—Deep Static | | | | | | | | |
| Repetitive Squat (Dynamic) | | | | | | | | |
| *Static Work* | | | | | | | | |
| Sitting | | | | | | | | |
| Standing* | | | | | | | | |
| Driving* | | | | | | | | |
| *Ambulation* | | | | | | | | |
| Walking | | | | | | | | |
| Stairs | | | | | | | | |
| Ladders | | | | | | | | |
| Balance | | | | | | | | |
| Uneven Terrain* | | | | | | | | |
| Running* | | | | | | | | |
| *Coordination* | | | | | | | | |
| R Upper Extremity | | | | | | | | |
| L Upper Extremity | | | | | | | | |
| R Cumulative Work* | | | | | | | | |
| L Cumulative Work* | | | | | | | | |
| *Aerobic Capacity* | | | | | | | | |
| Step Tests* | | | | | | | | |

\*Optional.
*Abbreviations:* CG, center of gravity; L, left; R, right.
*Source:* Copyright © 1988, Susan J. Isernhagen.

Frequency indicates % of an 8-hour day.

**Exhibit 14–3** Functional Capacity Evaluation: Summary Report

## FUNCTIONAL CAPACITY EVALUATION: SUMMARY REPORT

FORM C

Name:

Address:

Date of Birth:

Referral Source:

Attending Physician:

Dates of Test:

Description of Tests Done:

Consistency of Performance:

Objective Observations:

Cooperation:

Safety:

**Exhibit 14-3 Continued**

Quality of movements:

Outstanding physical features:

Physical Return-to-Work Options Explored:

Work Strengths Compared to Job Description:

Limitations, re Job Description:

Recommendations:

*Source:* Copyright © 1988, Susan J. Isernhagen.

**Table 14-5** Functional Capacity Evaluation Test

| Evaluation Item | Objective | Movement parameters | Equipment | Rating |
|---|---|---|---|---|
| Lifting | | | | |
| A. Floor to center of gravity (CG)—vertical | Lift weight receptacle (WR) from floor to client CG | Begin at CG shelf height (adjusted to client), lower WR to floor and return—5 repetitions | Weight receptacle approximately 12 in × 12 in × 12 in<br>Weights<br>Adjustable shelving | Five safe lifts within 1 min<br>Maximum poundage |
| B. CG to eye level—vertical | Life weight receptacle (WR) from CG level to eye level | Begin at CG level, raise WR to shelf height at eye level (adjusted) and return—5 repetitions | As above | Five safe lifts within 1 min<br>Maximum poundage |
| C. CG to CG—horizontal | Life WR from CG level, move horizontally 4 ft and place at same level | Begin at CG shelf height, pivot, move 4 ft, touch WR to shelf, return—5 repetitions | As above | Five safe lifts within 1 min<br>Maximum poundage |
| Carry | | | | |
| A. Bilateral front | Carry WR in two-handed manner at CG level | Begin with WR at CG level, carry 50 ft | As above | Safe carry<br>Maximum poundage |
| B. Right-handed | Carry WR in one-handed manner at thigh level | Begin with WR at thigh level shelf height, carry 50 ft using right upper extremity | One-handled weight receptacle: 10 in × 12 in × 12 in with handle in middle | Safe carry<br>Maximum poundage |
| C. Left-handed | As above | As above; use left upper extremity | As above | Safe carry<br>Maximum poundage |

**Table 14–5** Continued

| Evaluation Item | Objective | Movement parameters | Equipment | Rating |
|---|---|---|---|---|
| Push/Pull | | | | |
| A. Push | Move weight sled horizontally 10 ft | Handle adjusted to chest level on weight sled, using smooth force, push 10 ft | Weight sled with adjustable push/pull bar Scale—dynamometer | At maximum load, use scale to measure poundage needed to initiate movement |
| B. Pull | As above | As above; pull 10 ft | As above | As above |
| Handgrip | | | | |
| A. Right | Find maximum isometric handgrip strength Find optimum handgrip span | With elbow at 90 degrees use maximum grip force to squeeze handle; use 1st, 2nd, 3rd and 4th positions | Jamar-type dynamometer Tape measure | Average of three trials at each position Document greatest force and optimum grip span circumference |
| Elevated Work Neck/back in extension position | Work with upper extremities in elevated position while standing | Place assembly objects on shelf at worker eye level. Assemble items—5 min | Simple assembly activity Adjustable workbench | Amount of time elevated position is tolerated—5-min maximum |
| Lowered or Forward work Neck/back in flexion position | Work in the position of hip and trunk flexion<br>• while standing<br>• while sitting | Stand with hips/trunk flexed to 20 degrees | Simple assembly activity Adjustable workbench | Amount of time flexed position is tolerated—5-min maximum |
| A. Standing | | | | |
| B. Sitting | | Sit with hips flexed at 110 degrees | As above and stool | As above |

**Table 14–5** Continued

| Evaluation Item | Objective | Movement parameters | Equipment | Rating |
|---|---|---|---|---|
| Unweighted Rotation<br>A. Standing<br>B. Sitting | Move through diagonal and horizontal rotation patterns<br>• while standing<br>• while sitting | Move through three patterns:<br>1. Full right to left and reverse<br>2. Low right to high left<br>3. Low left to high right<br>10 cycles each | Empty, light-weight boxes, 12 in × 12 in × 12 in<br>Adjustable shelving<br>Straight-back chair without arms | Functional movement through full ranges of three motions, 10 cycles each |
| Crawl | Crawl on all fours | Reciprocal crawl 30 ft | Carpeted floor | Number of feet crawled, maximum 30 ft |
| Kneel | Kneel in upright position | Knees flexed to 90 degrees<br>Hips extended—1 min | Carpeted floor | Time able to maintain upright kneel—1 min |
| Sustained Crouch | Maintain full knee and hip flexion position | Position in flexed crouch—maintain 1 min | Rail for balance | Time crouch is tolerated—maximum 1 min |
| Repetitive Squat | Flex hips and knees fully to touch hands to floor | Stand, squat, return to stand—maximum 1 min minimum 10 repetitions | Rail for balance | Time squat is tolerated—maximum 1 min |
| Stair ambulation | Ascend and descend steps | Ascend and descend a flight of stairs until 100 have been traversed | Flight of stairs | Number of steps able to be ascended and descended, 100 maximum |
| Balance and stabilization | Step through lattice of squares while walking, demonstrating balance | Walk 40 ft with correct foot placement in squares, both diagonal and line pattern | Thin lattice of 18-in squares 10 ft long, 2 ft wide | Walks in two patterns 40 ft each, making less than two errors each |

**Table 14-5** Continued

| Evaluation Item | Objective | Movement parameters | Equipment | Rating |
| --- | --- | --- | --- | --- |
| Sitting | Sit in chair while performing table activity | Sit, weight shifting allowed, 30 min while performing hand tests | Standard armless, padded chair with straight back Supplemental items: lumbar roll, armrests | Sit functionally, 30 min maximum |
| Upper Extremity Coordination Right Left | Perform designated fine and gross motor coordination tests | Follow directions on hand coordination tests | As specified by hand coordination test | Scoring per test directions |
| Optional Functional Capacity Evaluation Test Items Walking Running Standing Outdoor Activity—Uneven Terrain Driving Step Tests Upper Extremity Cumulative Work | | | | |

*Source:* Copyright © 1988, Susan J. Isernhagen.

*Functional Capacity Evaluation* 159

**Figure 14–1** Example of FCE test item. Hand coordination tests are important in documenting fine motor coordination and hand-wrist position tolerances. Sitting also can be evaluated, both for postural adaptations and sitting tolerance.

---

The purpose of FCE is to define functional abilities and limitations only in the context of safe, productive work tasks. Physical conditions may exist that may predict dysfunction and they need to be identified. However, experience has shown that they can be found primarily by testing the specific activity, not by predicting function on the basis of a clinical examination.

**Example:** Robert Jones, a truck driver, has a knee injury. Preliminary examination shows:

- forward head positioning
- increased thoracic kyphosis
- increased lumbar lordosis

**Figure 14–2** Example of FCE test item. Dynamic pushing tests allow measurement of the amount of force able to be produced. They also facilitate evaluation of body mechanics and the strength of the upper extremity, trunk, and lower extremity components of the push.

---

- a leg-length shortness discrepancy of 0.5 in on the right
- decreased trunk motions by approximately 30% of each motion
- right knee flexion range of 0 to 105 degrees

Despite these findings, he tested functionally safe and able on all FCE items. The only exception was lift from the floor, which has to be modified with a 12-in riser to allow for loss of knee motion. Return to work was accomplished.

In looking at function, the positive functional aspects of the person must be accentuated. These abilities enable a return to work or to other functional activities. The limitations described will be only for the purpose of prevention of reinjury. The referral sources for FCE, and the injured worker himself, will be more interested in what is right about the activities, rather than what is wrong.

## Behavior Management

With an injured worker who is pain focused, it must be stated at the outset that pain and function are not synonymous. To illustrate, people with pain may be very functional in life. Conversely, a person could perform an unsafe procedure and actually sustain an injury before pain becomes noticeable.

The worker can be encouraged to report any fear or pain so it can be documented and discussed. It should be understood, however, that fear or pain will not in itself limit the testing. The encouragement toward maximal function by the therapist is strengthened by explanation of the task and assurance of its safety. There also should be a reassurance that if the task is physically unsafe, it will be stopped by the therapist and the physical reasons documented.

Often, maximal function and tolerable levels of discomfort are reached simultaneously. But, if a client begins to exceed safe activity levels, the therapist must stop the test and clearly explain to the client why activities beyond that level may be injurious. This should be documented as a "physical limitation." If this is not the case, however, and the client stops before maximal function is reached, the activity's maximum test results should be clearly designated as "client self-limitation."

## Malingering Management

Managing a patient who is not putting forth full effort is difficult. The therapist should be prepared to deal with it objectively.

When activities or motions are stopped short and no physical cause is seen, the therapist must be ready to confront the person with that information. It can be done in a very positive, objective, and nonjudgmental manner.

"Labeling" the patient (eg, as a malingerer) is not necessary. Merely stating the inconsistencies or lack of physical effort signs will be enough for others to draw their own conclusions.

### Encouragement of Self-Determination

A person who has gone through a process of injury, removal from work, and treatment to heal the injury can feel passive in the process by the time he or she reaches an FCE level. Lack of knowledge of physical capability perpetuates the passive role.

In FCE, the person is taken through a test that clearly shows abilities and physical limitations that are present. It aids the person in knowing that he or she is going to be fully capable of doing certain activities both at home and at work. Also, the objective expression of physical limitations will indicate to the person that there are limits to the activity. The limits will be clearly stated and understandable to the person. Once the person knows what could cause reinjury, he or she can take the responsibility of avoiding that activity both at work and at home.

## REPORT FORMATS

The report of the functional capacity evaluation is almost as important as the test. If the report is written in strongly technical or medical language, then employers, patients, and vocational consultants may not be able to fully understand the results. On the other hand, if medical specifics are not stated, then the medical practitioners and legal experts will not be able to interpret the recommendations. Therefore, the report should be written in concise, precise, and understandable language. Different formats that have been used in expressing functional capacity results follow.

### Checklist

A checklist is efficient for time use of the therapist. Examples are rating systems based on a scale of 1 to 5, or a judgment of "able" or "not able." This format allows the employer, insurance company, or physician to see with ease what the person can do and where his or her limitations are.

The drawback of this type of recording lies in its inability to verify, at a later date, how the results were determined. The physician and employer may not be able to realize the full professional skills that were used in making an accurate determination.

## Narrative Form

A narrative format provides more opportunity to substantiate the conclusions. It is easier for the employer and physician to understand and implement.

Narrative reports are often long and time consuming to write. They are particularly beneficial when used in court or in a specific return-to-work determination. The most useful narrative reports also include a brief summary statement.

## Workers' Compensation State Return-to-Work Forms

Workers' compensation forms (ie, the forms signed by the physician to allow a worker to go back to work) contain a listing of many work activities. There are many categories mentioned, but there is very little definition of terms.

Specifically in the lifting category, there often is no differentiation among types of lifts. For most people the amount of weight that can be lifted from the floor is different from the amount of weight that can be lifted from table height. However, because the lift is not defined, estimates, rather than specifically tested and documented items, tend to be recorded. The information has value, however, as most state workers' compensation systems find the data easy to understand and compare to a job description.

It is possible to take the results of an FCE and apply them in filling out a state workers' compensation form more specifically. The FCE would define the type of lifts that were tested, and these can be differentiated on this form. State workers' compensation professionals are more than happy to see detailed information put on a general form. The therapist can feel confident in adding more information than is requested.

## Combination of Best Features

A combination of the previously discussed reports has led to the development of a format that is not only easy to read but quite specific in stating abilities and limitations. It combines the checklist format with added comments on physical abilities and limitations, as well as other parameters of the testing situation. This allows the therapist to use a format that is not lengthy and to make many positive statements about the injured worker that go beyond the usual checklist format (see Exhibits 14-1, 14-2, and 14-3).

## CASE STUDY—FUNCTIONAL CAPACITY EVALUATION

A case study of an FCE performed on a work-injured nurse is presented in Appendix 14-A. A description of the patient, her abilities, and her limitations are given in a clear yet brief format, using Forms A, B, and C (Exhibits 14-1, 14-2, and 14-3, *infra*). Return-to-work recommendations are provided for the physician, employer, rehabilitation consultant, insurance company, and participating employee/patient.

**Note to Evaluator:** Information to complete Forms A and B is gained from the FCE. Form C (summary report) becomes a compilation of information derived from Forms A and B. For the most effective use of Forms A and C, have them keyed onto a word processor and allow for proper spacing as results are typed in final form.

## ADJUNCTS TO FUNCTIONAL CAPACITY EVALUATION

Because of the need to specifically define many medical parameters, FCE equipment often is used as an adjunct to professional evaluation. Four examples of adjuncts to FCE that are instrumentation-equipment oriented follow.

### Isokinetic Testing

Isokinetics are well accepted as an advanced muscle function evaluation tool and treatment mode. The ability to set the speed of a movement while matching maximum resistance adds substantially to physical therapy evaluation parameters. Isokinetic testing is important in identifying specific muscle group weaknesses and areas of dysfunction within stabilized arcs of movement. Comparisons of agonist and antagonist also are helpful.

In the past years, several firms have developed isokinetic testing devices. There are differences in stabilization methods, measurement methods, and calibration verification. Some equipment can be adapted for strong stabilization with exact measurement of body movements. Other equipment is designed to use freer movement, allowing "functional" patterns, but some ability to verify specific muscle groups being measured is lost.

If the therapist administering an FCE can identify areas of weakness or instability, there is a benefit to the addition of specific testing of the weak muscle components. The identification and validation of areas of deficiency through accurately recorded testing is very helpful, not only in producing an accurate evaluation but also in making recommendations for further treatment.

## Isometric Testing

Isometric testing has been done extensively because it is easy to use, has accurate measurement methods, and has a body of research.

Testing done in several lifting, pushing, and pulling positions allows an examiner to know how much force can be generated in a certain direction. This is valuable information in correlating with actual functional movements.

The weaknesses in this testing, however, must be understood when correlations to function are made. Isometric testing is in a functional direction, yet the ability of the body to "set" itself before doing a maximum contraction removes isometrics from the realm of functional movement. True function requires a beginning and an end, an arc of motion, momentum, and concentric or eccentric contractions.

If, in the future, FCE is shown to be highly correlated with isometric testing, it is possible that both could be used to validate a person's functional ability.

## Motion Analysis

Since the advent of computerization, a new technology has emerged that allows movement to be measured by evaluating motion of computerized points on the body during functional activities. For example, a knee joint, hip joint, and ankle joint can be tracked through a functional walking or jumping activity. This can then be correlated with specific muscle groups and lead to a better understanding of the relationship of the muscles, the joints, and the motion. This is amenable to use in validating FCEs. Further study of muscle groups involved over those points might be able to provide better information in designing a restorative program.

## Cardiopulmonary Evaluation

In FCE, heart rate, blood pressure, and respiratory rate can be measured at rest and with activity, and then can be analyzed to note the changes that occur with activity. In many cases, maximum permissible limits must be set for a person and then monitored to ensure the limits are not exceeded during FCE.

Because accurate studies have not been done on all functional work motions using cardiopulmonary screening, more research in this area is desirable. Evaluation of oxygen use during activity may lead to prediction of the endurance rates of people. Recording of changes in heart rate, blood pressure, or respiratory rate may also lead to recommendations for conditioning.

## LEGAL IMPLICATIONS

Because the workers' compensation system or the personal liability system is very likely to involve litigation, the professional doing FCEs is more likely to be involved in legal action. Therefore, it is critical that all FCEs be done with objectivity and use of full professional skills, with the thought that records may be subpoenaed for a courtroom situation. In addition, the evaluator-therapist also may be called on to legally defend the recommendations and outcomes of the FCE.

It is very important that the evaluator be objective and testify on physical signs and symptoms, rather than subjective merits of any particular case. The legal merits of the case will be adjudicated in court. The therapist's report will merely be used as information to aid in determination. The therapist should not judge the motivations of people. The reports should reflect physical abilities and limitations and other objective return-to-work parameters.

### Courtroom Testimony and Depositions

When an evaluator is called to report on the results of an FCE, it will most likely be as an expert witness. He or she will not be merely relating what was seen; he or she will be asked to give a professional interpretation of the evaluation and the recommendations. Therefore, the attorneys involved will establish the professional credibility of the evaluator at the outset. If it is a therapist, the type of credentials, type of schooling, and type of experience will be determined before any testimony is taken.

In many cases, more than one medical professional is called. For instance, two physicians and the therapist may all be called to testify on the same case. The FCE evaluator need not feel intimidated by the presence of other medical witnesses. A quality FCE will have been objectively done and thoroughly documented, and the evaluator will be merely presenting the information on what happened during the testing situation and the functional outcome.

In a deposition, the judge and jury will not be present. Two attorneys and a recorder will be present. In a deposition an attorney is more likely to be aggressive and manipulative and to pressure the witness. The evaluator can prepare for this event by being aware of his or her own credentials and professional work with the patient. The evaluator need not answer questions that are beyond the professional scope of his or her qualifications or the scope of the actual test.

In a deposition, the written FCE report will also have been called as part of the evidence. Therefore, the therapist will have the report for reference in thorough explanation of what was tested and what was found.

Whether in deposition or courtroom testimony, an FCE evaluator who has done an objective, comprehensive test of a patient will be successful if a calm presentation and professional attitude are maintained.

# Appendix 14-A
# Functional Capacity Evaluation Case Study

## FUNCTIONAL CAPACITY EVALUATION: PATIENT DATA
*History*

Form A

Date: 7-1-88

Name: Jessica Lorayne

Address: 4444 Aura Avenue
Duluth, MN 55000

Telephone No.: (121) 910-2222

Referral Source: Laura Black,
Rehabilitation Consultant

Physician: Dennis Dean, MD

Diagnosis: Chronic Low Back Strain

Date of Birth: 8-3-47

Date of Injury: 9-19-87

Type of Injury: Sudden strain caused during patient transfer. Back in flexion posture.

Total Time Off Work: 10 months

Occupation: Registered Nurse

Employer: Simms Nursing Home

Insurance Company: Norwood Mutual

Pertinent Surgery: None

Previous Treatment:
Physical therapy, October and November 1987. Treatment consisted of heat, massage, and extension exercises. Discharged with residual symptoms.

Current Medications:
Valium discontinued 6-1-88.

Patient Report:
- Functional Level: Inability to tolerate sitting, standing, or walking more than 15 minutes at a time. Does light housework but no other work. No aerobic exercises.
- Pain Level: On a 0 to 10 scale, current level: 5. Level after most activity: 8.

Work Status:
On disability leave; workers' compensation benefits being paid.

## PHYSICAL ASSESSMENT SUMMARY

Blood Pressure: 140/88

Resting Pulse Rate: 92

Respiration Rate: 18

Gait: Slow and guarded with little trunk movement. No other abnormalities.

Posture: No major abnormalities. Slight increased lumbar lordosis, thoracic kyphosis, and forward head.

Coordination: No apparent deficiencies in coordination.

Balance: Patient reports balance loss on stairs and uneven ground. She was tested with alternate one-leg stands. She was able to maintain only 2-second stand in test of each lower extremity. Weakness in hip stabilization for this type of balance bilaterally.

Movement Characteristics: Coordinated movements all done in slow, guarded manner.

Strength: Dynamic muscle testing through mid-range, 24 to 40 degrees of movement. Patient able to tolerate moderate resistance for all extremity muscle groups. In active sit-ups and back extensions done against gravity, she could tolerate no resistance, although able to move through partial range against gravity.

Joint Range of Motion:

*Upper Extremity*: Both right and left upper extremities within normal limits.

*Lower Extremity*: Joints within normal limits with the exception of hamstring tightness. Straight leg raising to 45 degrees bilaterally.

*Cervical*: Flexion, extension, and lateral flexion within normal limits. Rotation limited to 70 degrees bilaterally. Discomfort at the ends of the ranges.

*Trunk*: Trunk flexion to 65 degrees while in standing position. Discomfort reported from 30 to 65 degrees of movement.

*Side Bending*: 20 degrees bilaterally. Discomfort at end of range.

*Rotation*: Slow, guarded movement in rotation, which is limited to 20 degrees bilaterally. Patient states she avoids trunk bending and movement because of discomfort.

Comments: Patient guards against trunk movement in all activities. Discomfort at ends of ranges of trunk and cervical motion is present before full range has been tested. She moves in a nonenergetic manner.

Physical limitations noted were trunk weakness, trunk decreased motion, discomfort during ends of neck and trunk motions, and mild hip instability in gait and standing.

High pulse rate correlates with low endurance level and low aerobic capacity.

*Source:* Copyright © 1988, Susan J. Isernhagen

## FUNCTIONAL CAPACITY EVALUATION

**FORM B**

Client: Jessica Lorayne
Date: 7-1-88

| Item | 0%-5% | 6%-25% | 26%-50% | 51%-75% | 76%-100% | Restrictions | Comments |
|---|---|---|---|---|---|---|---|
| **Weight Capacity in Pounds** | | | | | | | |
| Lift: Floor to CG | 15# | 0 | 0 | 0 | 0 | Hip stabilizer and quadriceps weakness limitation | Cannot be done repetitiously |
| CG to eye level | 25# | 10# | 0 | 0 | 0 | Back extension limitation | Should not reach higher than eye level |
| Horizontal | 30# | 15# | 10 | 0 | 0 | Low endurance | Smooth movements |
| Push: Dynamic | 40# | 15# | 10# | 0 | 0 | Low endurance | Should move slowly |
| Pull: Dynamic | 50# | 15# | 10# | 0 | 0 | Low endurance | Should move slowly |
| Carry: R | 15# | 0 | 0 | 0 | 0 | Trunk stabilizers weak | |
| L | 10# | 0 | 0 | 0 | 0 | | |
| Front—at CG | 35# | 25# | 10# | 10 | 10 | Guarded trunk position | Should walk slowly |
| Handgrip: R | 55# | 30# | 10# | 10 | 0 | | Must keep back erect |
| L | 52# | 30# | 10# | 10 | 0 | | |
| **Flexibility/Positional** | | | | | | | |
| Elevated Work | | | | | | Back extension limitation | Must self-pace any reaching |
| **Trunk Flexion Work** | | | | | | | |
| Sitting | | | | | x | Flexes at hips; keeps back extended | Should alternate positions |
| Standing | x | | | | | Tolerates trunk flexion poorly | |
| **Unweighted Rotation** | | | | | | | |
| Sitting | x | | | | | Limitation in trunk motion | Should use swivel chair |
| Standing | x | | | | | | Should pivot at feet |
| Crawl | x | | | | | Weak back extensors | Limit or avoid crawling |

172  Work Injury

| | | | |
|---|---|---|---|
| Kneel | x | Unstable at hips, poor balance | Should not crouch |
| Crouch—Deep Static | 0 | Hip stabilizer and quad weakness | Minimize squatting |
| Repetitive Squat (Dynamic) | x | | |
| *Static Work* | | | |
| Sitting | x | | Should alternate positions |
| Standing* | x | | Should alternate positions |
| Driving* Not tested | | | |
| *Ambulation* | | | |
| Walking | x | Deconditioning prevents speed | Capacity moderate but should walk slowly |
| Stairs | x | Hip and quad weakness | Should self-pace and use rail or ladder for support |
| Ladders | x | Hip and quad weakness | |
| Balance | x | Weak hip stabilization | Should do activities requiring high balance skills |
| Uneven Terrain* Not tested | | | Not tested. |
| Running* Not tested | | | |
| *Coordination* | | | |
| R Upper Extremity | x | | Excellent aptitude |
| L Upper Extremity | x | | Excellent aptitude |
| R Cumulative Work* | | | Not tested. |
| L Cumulative Work* | | | Not tested. |
| *Aerobic Capacity* *Step Tests** | | | |

Frequency indicates % of an 8-hour day.

*Abbreviations:* CG, center of gravity; L, left; R, right.
*Optional.
*Source:* Copyright © 1988, Susan J. Isernhagen.

## FUNCTIONAL CAPACITY EVALUATION: SUMMARY REPORT

**FORM C**

Name: Jessica Lorayne

Address: 4444 Aura Avenue, Duluth, MN 55000

Date of Birth: 8-3-47

Referral Source: Laura Black, Rehabilitation Consultant

Attending Physician: Dennis Dean, MD

Dates of Test: June 30, July 1, 1988

*Description of Tests Done:* Ms. Lorayne participated in all functional capacity evaluation tests. In the optional category she also took the walking and standing tolerance evaluations.

*Consistency of Performance:* Within each test item, client demonstrated consistent performance for the repetitions or time limit tested. However, there was a noted improvement in performance on the second day of testing. She performed with some self-limitations on the first day, but by the second day of testing stated she felt much more comfortable in understanding her true capabilities. The performance on the second day is consistent with performance should she return to work or increase daily activities at home. These scores are documented on Form B.

Her performance is consistent with a patient who has been self-limited because of fear of reinjury. The second day's performance indicated that she had overcome the fear and is able to perform at a higher functional level.

Objective Observations:

*Cooperation:* Although client was cooperative for the two days of testing, she was more eager to participate on the second day. The items that caused her the greatest fear were the lifting tasks. But, when they were repeated in the testing, she did not appear fearful and was pleased with her scores.

*Safety:* Although she moves with a slow, guarded movement, all her motions are safe. She used good body mechanics throughout testing.

*Quality of Movements:* She exhibited smooth motions and had excellent hand coordination. Balance testing stressed her, but she slowed her pace to ensure adequate balance during this type of testing. All other movements were coordinated.

*Outstanding Physical Features:* Decreased trunk motion was typical in all her activities. Although this did not prevent good functional scores on test items, it is a feature that should be investigated further for remediation.

*Physical Return-to-Work Options Explored:* Her rehabilitation consultant, Laura Black, stated that Ms. Lorayne could rejoin the Simms Nursing Home as a staff member in either an RN position or a night supervisor position. Therefore, this test addressed capacities needed for both positions.

*Work Strengths Compared to Job Description:* Because the job description of unit RN requires a significant number of patient transfers, Ms. Lorayne would not meet the lifting requirements for this job description.

The night supervisor position would not require patient transfers, and would mainly consist of recordkeeping, light patient care, and passing medications. Functional actions involved are sitting, standing, walking, moderate flexing during light-duty care, and supervisory and counseling activities that could be carried out in a self-selected position.

The physical requirements of the night supervisor job appear to be within the functional capacity abilities of the client. Therefore, there is a good match for physical abilities and a job description of night supervisor nurse as a result of this Functional Capacities Evaluation.

*Limitations, re Job Description:* If supervisory nurse job description is adhered to, there should be no physical restrictions. It will be helpful, however, if self-pacing and frequent change of positions is allowed. Client needs to alternate sitting, standing, and walking so that no one activity is done for prolonged periods. She fully understands her ability to self-pace and change positions and should be able to accomplish this at work.

If lifting is required, she will be limited to the pounds of lifting indicated in Form B. This eliminates patient transfers but not wheel chair pushing.

*Recommendations:*

1. Continuation of return-to-work process. Basic physical capacity results match nurse supervisor job description with exception of endurance.
2. Evaluation of trunk mobility, trunk strength, and hip stabilization.
3. Evaluation for work hardening or fitness program to increase aerobic capacity, lower extremity extensor strength and endurance, back mobility and stability, overall efficiency and speed of movement, and nursing activities.

*Source:* Copyright © 1988, Susan J. Isernhagen.

# 1. Disability Determinations

*Jack Casper*

It is important for all physical and occupational therapists performing functional capacity tasks to realize fully the implications of the report in a courtroom. Functional capacity evaluation (FCE) reports are used primarily to determine base line functioning of an injured worker and to make recommendations regarding services or activities that may increase functioning. Also, they are typically used to identify the level of physical functioning so that a worker is not returned to a job that is beyond his or her capabilities. The FCE evaluator normally works with a medical or vocational rehabilitation consultant who coordinates a plan for return to work.

Because rehabilitation consultants consistently have direct work experience in returning disabled persons to work, they often are called on to testify at legal proceedings regarding a person's ability to return to work. Through careful analysis of previous work history, transferable skills, and residual functional capacity, a rehabilitation consultant is in a position to testify regarding whether a person might be totally disabled from gainful employment, or whether someone has lost earning capacity because of an injury or condition. In testimony situations regarding these legal matters, the rehabilitation consultant puts on a different hat and becomes known as a "Vocational Expert."

In determining total or partial disability from gainful employment, the single most important piece of evidence a vocational expert will use is a written assessment of the claimant's residual functional capacity (RFC). Very often, the RFC assessment comes from an FCE assessment center. In the author's experience, determinations made by comprehensive FCEs are typically approved by physicians. The physician has referred the person for the evaluation because there is a need for functional capacities to be objectively measured so that the baseline functional capacities will be known. The physician certainly has no reason to disbelieve the results, and, typically without question, considers the FCE report objective medical evidence that he or she would most certainly cosign.

In a legal situation, when the degree of a disability must be adjudicated, vocational experts, judges, and attorneys know that a formal opinion of the claimant's RFC is more important in determining disability than a huge stack of medical documents, physician's letters, and so forth, that do not include a judgment of RFC. The court needs to know what a person is physically capable of doing, and for how long, to make any reasonable determination of disability.

Following the rationale in the above statements, a significant conclusion can be made: in numerous disability cases heard in the legal system, a physical or occupational therapist's judgment of a person's functional capacity may be the most important factor in the final outcome of the case. In other words, the judgment could very well mean the difference between a huge financial settlement (or lifelong disability payments) or no monetary award at all.

## SOCIAL SECURITY DISABILITY HEARINGS

To demonstrate how disability hearings work, the example of Social Security disability can be used. If a worker has paid into the Social Security system for a required period of time and has an injury or condition that prevents him or her from performing any substantial gainful employment, that person is entitled to receive Social Security payments from the date of the disability until such time as the person might again be able to partake in gainful employment activity. Typically, if Social Security disability payments are granted, many of those receiving these payments do not return to work for the rest of their lives (although a certain percentage do eventually return to some type of gainful employment). If a person files written requests for Social Security disability payments and is denied, the person may then request a hearing on the matter conducted by an administrative law judge for a judicial decision regarding his or her claim.

At these hearings, the Social Security Administration will usually retain the services of a Vocational Expert to testify whether or not, based on the person's disability, he or she is able to perform one or more jobs found in substantial numbers in the national or regional economy.

At the hearing, the judge will thoroughly examine all medical evidence submitted to the Social Security Administration and will question the claimant regarding his or her condition, the activities he or she is able to perform, and so forth. Normally, the claimant will have retained the services of an attorney who also can ask pertinent questions of the claimant and cross-examine the vocational expert.

After the judge and attorney have questioned the claimant and reviewed all pertinent medical evidence, the judge will then pose one or more hypothetical questions to the Vocational Expert. A hypothetical question is developed by the judge after taking all the facts and subjective reports into account. A typical hypothetical question to a Vocational Expert might be the following:

> Mr. Casper, given a man 50 years of age, with a tenth grade education, and a vocational history as stated, and, assuming the following physical capabilities: ability to lift up to 20 lb on an occasional basis and 10 lb on a frequent basis, only occasionally bending and stooping, ability to walk four hours in an eight-hour workday, to sit four hours in an eight-hour workday, and to stand two hours in an eight-hour workday.—Mr. Casper, would you give me your opinion as to whether there are any jobs in the national or regional economy that the claimant can work, and, if so, what are your reasons for stating such?

Taking all the judge's hypothetical factors into account (and noting that the judge has based the questions on his or her interpretation of all information presented), the Vocational Expert then responds to that question in an impartial manner based on objective vocational rules.

Functional capacity evaluation documents are used often in various types of disability hearings. Typically, these are the definitive documents on which RFC is based and, therefore, pervade the entire case in terms of the questions that might be asked of experts, and in essence, are used in making a final decision regarding a disability award in a case.

## STANDARDIZED JOB TRAITS

In terms of the identification and classification of jobs, there are two widely accepted sources that are all pervasive in legal determinations. The US Department of Labor[1] and the Social Security Administration classify jobs primarily in terms of exertional level, skill level, and other categories. All jobs in the US economy fit into one of five levels of exertion according to the US Department of Labor and the Social Security Administration. These levels are:

1. *Sedentary Work*—Sedentary work typically requires sitting for six hours out of an eight-hour workday, lifting no more than 10 lb on an occasional basis, with possible frequent lifting of small objects weighing less than 10 lb (eg, files, dockets, ledgers). Many clerical, cashiering, managerial, assembly, and desk work jobs are classified under this category.
2. *Light Work*—Light work is defined as lifting no more than 20 lb on an occasional basis and up to 10 lb on a more frequent basis. Light work typically requires standing and walking for six hours out of an eight-hour workday. Certain light jobs may require continuous sitting, however, and would also entail the consistent use of either hand or foot controls (eg, crane operator, certain equipment operation jobs).
3. *Medium Work*—Medium work typically requires lifting up to a maximum of 50 lb on an occasional basis and up to 25 lb on a frequent basis. It also is generally accepted that medium work necessitates the worker being on his or her feet six hours out of an eight-hour workday.
4. *Heavy Work*—In this category, maximum lifting of 100 lb is expected, with more frequent lifting of weights up to 50 lb. Standing and walking six hours out of an eight-hour workday also is included in this category.
5. *Very Heavy Work*—Maximum lifting of more than 100 lb is required in this type of work, with more frequent lifting of weights up to 50 lb. Standing and walking for the majority of the workday also is included.

Another important factor to be dealt with is skill level. Jobs are typically categorized into one of three skill categories: unskilled, semiskilled, and skilled. If a person has previously performed semiskilled or skilled work, that person generally has skills that transfer to other types of occupations.

The discussion of skill levels is comprehensive. There are many examples of occupations that are light, medium, or heavy in nature and could be considered as semiskilled or skilled work. However, skills may not transfer to sedentary types of occupations; and, in the case of skilled medium work, skills may not transfer to light or sedentary occupations.

If a person had previously performed medium-duty skilled work and had a condition to the extent where he or she could now only perform sedentary work, chances are he or she could not use the skills in transferring to a sedentary type of occupation. The person then would be relegated to sedentary unskilled work. (Sedentary unskilled work is a very minimal category in terms of the number of jobs available.) A person with this job history and restrictions might well be determined to be disabled totally and completely because of this profile.

## ALTERNATE SIT/STAND JOBS

Not all jobs fit into the above five categories exactly. Especially in terms of sedentary and light work, there are jobs for which a person might be able to alternately sit and stand at will. These types of positions typically encompass skilled and managerial type positions where a person is supervising or managing. Also, certain types of cashiering and receptionist jobs would fall into an alternate sit/stand category.

There is only a small job base for someone who must alternate sitting and standing positions unless that person previously performed managerial or supervisory types of work.

## DISCUSSION

The following simplified examples indicate the relationship between restrictions and work:

**Example 1:** A person who is limited to lifting no more than 10 lb and could sit only two to three hours in an eight-hour workday would probably be judged to be totally disabled unless that person had managerial, supervising, or cashiering types of skills. The person is limited to sedentary work by virtue of the lift restriction; however, he or she cannot perform sedentary work because of the sitting restriction. This results in an elimination from all job categories.

**Example 2:** Lifting restriction of 20 lb; sit two hours, stand two hours, and walk four hours.—this person typically would not be classified as disabled. His or her RFC allows a full range of light work to be done by virtue of lift restrictions and stand/walk restrictions.

**Example 3:** Lift 25 lb; stand two hours, walk two hours, and sit five hours; claimant also has arthritis in both hands. Judgment: probably disabled. Although the person has lift restrictions allowing light work, standing and walking restrictions eliminate this category. Five hours of sitting is probably sufficient to allow sedentary work; however, because most sitting types of job activities include fine and gross manual manipulation, sedentary work is eliminated because of the arthritic hands.

**Example 4:** Lift 10 lb; sit two to three hours, stand two to three hours, and walk two hours; person has previous supervisory experience. Judgment: not disabled. Although it would appear that this person is precluded from all work categories, supervisory experience would typically allow an alternate sit/stand type of job.

**Example 5:** Lift 50 lb; sit six hours, stand one hour, and walk three hours; manual labor background; aged 55. Judgment: disabled. Although lift restriction allows medium work, standing and walking tolerance preclude medium and light work. Sedentary work category is open. However, a person of this age is not expected to make a successful adjustment to a different type of work (sedentary activities are much different from manual laboring activities).

As can be seen from the above simplified examples, many nuances enter into the judgment of disability. Also, there is a whole range between total disability and zero disability, when a person might not be totally disabled but has lost earning capacity because of his or her injury and inability to perform a full range of work.

The bottom line for all disability judgments is the person's RFC. All other analyses (eg, skill level, skill transferability) must first take into account the physical abilities of the person.

In summary, an FCE is typically the starting point, or bottom line, in adjudicating disabilities. The work of the therapist in judging functional capacity is extremely important and, in many instances, can completely alter a person's way of life if some type of legal adjudication of disability is in the picture.

---

**REFERENCES**

1. Appendix C, physical demands, in *The Classification of Jobs According to Worker Trait Factors: Addendum of Occupational Titles, 1977*. Roswell, Ga, Vocational Services Bureau, 1978.

# 2. Cost Savings in Four Cases

*Margot Miller*

The role of Functional Capacities Assessment (FCA) (copyright Polinsky Medical Rehabilitation Center, Duluth, Minn) in returning workers to work has grown dramatically. As a result of accurately facilitating return to safe work, it is a cost-saving measure for employers and insurance companies.

Referral patterns and cost savings with FCA have been studied for the past 5 years. Although early referral has been most effective in early, safe return to work, the FCA is also often used in chronic situations. It is in those cases that cost savings is most dramatic.

Kelsey and White[1] document that after being off work for 6 months continuously with low back pain, only 50% of injured workers will ever achieve employment again, only 25% will return to work after 1 year, and a negligible amount will be able to work after 2 years. The emphasis with FCA is to change these patterns through evaluation of 17 functional work items, thus effecting a safe return to work for the employee. As a result there has been a significant decrease in lost work time and in workers' compensation costs.

The following case studies illustrate the cost effectiveness of FCA. The workers' compensation cost estimates are based on the Minnesota Workers' Compensation Department of Labor and Industry 1985 figures. The projected additional time off work is based on a combination of Kelsey and White's[1] information and our overall experience in patterns in the workers' compensation system.

## CASE STUDY #1

A 38-year-old administrative clerk for the telephone company sustained a low back injury. She also had a history of bilateral knee dysfunction. The physician was hesitant to release her for work because of her overall physical problems. At the time of the FCA, she had been off work for 12 months and there were no plans for return. By completing the FCA and comparing the test results to the job analysis, however, it was found that her physical abilities were adequate for her job. The client had mild limitations in kneeling, stooping, stair climbing, ladder climbing, and stand up lift. Her job was primarily desk work, and the limitations did not interfere in returning to work. She had no limitations in the more important job tasks of prolonged sitting, overhead work, hand coordination, or level lifting.

The FCA cost was $456. Projecting a 12-month additional time off work if the FCA had not been used, a workers' compensation cost of $10,000 would have resulted. An additional cost for salary replacement would be $15,000. Projected FCA savings: $24,544.

### CASE STUDY #2

A 57-year-old police captain sustained a head wound resulting in a cranioplasty and accompanying physical limitations. The employer anticipated return to work for him but planned conservative placement at a lower level than the original position. The FCA testing, however, showed high ability levels, and the report was instrumental in facilitating client's return to work at full duty without restrictions. The client was capable of lifting and carrying 80 lb repetitively and performing other tasks (eg, climbing, bending, and stooping) that required strength and coordination. Additionally, he had no speech or memory problems.

The cost of the basic FCA was $432. Psychology and speech components were $378. Projecting workers' compensation cost by placing client at a lower level position for 8 years until retirement results in a difference of $71,136, with a total replacement salary of $96,000. Projected FCA savings: $166,326.

### CASE STUDY #3

An FCA was performed on a 47-year-old janitor for a city newspaper who sustained a lumbar disk injury as a result of lifting a file cabinet. A job analysis was supplied by the employer and the FCA results were compared to it. Client was capable of level lifting and carrying 50 lb, and lifting 40 lb from the floor. Decreased trunk flexibility limited forward bending. In this case, the client could perform only partial job duties, but the employer was willing to modify the job. The modified job included no lifting over 50 lb and no mopping, sweeping, or emptying deep waste bins. Therefore, the client returned to work 3 months earlier than expected. A follow-up FCA was performed 3 months after the initial FCA to determine if job duties could be increased. The follow-up FCA results showed capability to return to work at full duty. Client was then capable of level lifting and carrying 60 lb, and 50 lb in stand-up lift. No limitation was noted in forward bending, enabling him to perform all job duties.

Cost of the initial FCA was $480 and the follow-up FCA was $120. Projected workers' compensation cost for the 3 months the employer modified the job was $3,692.28, with a salary replacement cost of $6,000. Projected FCA savings: $9,092.28.

## CASE STUDY #4

A 29-year-old trimmer at a food preparation plant sustained a shoulder injury at work. Because of perceived pain during activity, the client doubted she could return to work. The physician was unsure of her abilities and referred her for an FCA. The FCA documented the client could safely lift and carry 15 lb. All lifting and carrying required on the job were below that poundage. Therapist interaction helped the client understand that with proper pacing and smooth movement patterns, waist-level activities could be performed safely and with no increase in perceived pain. Overhead work and repeated stooping were functionally limited but did not interfere in return to work, as all work tasks were at waist level. Comparing FCA results with the job analysis indicated she could return to work.

FCA cost was $480. At the time of the FCA she had been off work for 8 months. Projecting 8 months of additional time off without an FCA generated a workers' compensation cost of $5,743.68 and a salary replacement cost of $9,360. A permanent partial disability settlement is also projected, which would be $12,923 paid over a 72-week period. Projected FCA savings: $14,623.68, plus a disability savings of up to $12,923.

In all cases noted, restorative physical exercise suggestions were given to allow the injured worker to improve physical limitation areas even if they did not interfere with work.

In each of these cases the FCA was instrumental in an earlier return to work for the employee, as well as savings of thousands of dollars. The physician, employer, and employee can all benefit by clear-cut, objective assessment of functional work capacities.

---

**REFERENCE**

1. Kelsey JL, White AA III: Epidemiology and impact of low back pain. *Spine* 1980; 5:133–134.

# 3. Functional Capacities Assessment Research: The Relationship of Age and Gender to Functional Performance—Patients and Uninjured Subjects*

*Susan J. Isernhagen, Karen Mokros, Margot Miller, and Laurie Johnson*

## OVERVIEW

Extensive clinical testing of work-injured patients has revealed correlations of function with age and gender. One thousand six hundred and thirty-nine patients were evaluated with a standardized Functional Capacities Assessment (FCA) over a 5-year period. The results indicate that age and gender are significant factors in three functional areas: (1) the ability to move weight; (2) the ability to stair climb; and (3) the ability to move on a balance beam.

To determine if the same relationships were present in an uninjured working or active population, a limited study was performed. The relationships of age, gender, and functional ability were similar to the patient group.

Regarding age, the younger subjects did better on the FCA than the middle-aged subjects, and the middle-aged subjects did better than the advanced-aged subjects. Regarding gender, the female subjects generally performed less ably than their male counterparts, but this was less clear in the younger groups.

In the work setting, this may be an indication that older workers, female workers, and particularly older female workers may be at higher risk in functional work activities than their younger or male counterparts. This may lead to considerations of work or worksite modifications in maintaining safety of the employee. It also is an indication that there may be a stronger need in the older groups for restorative work hardening after injury.

---

*The authors performed this research study for FCA Network, Polinsky Medical Rehabilitation Center, Duluth, Minn.

## INTRODUCTION

The Functional Capacities Assessment (FCA) (copyright, Polinsky Medical Rehabilitation Center, 530 East Second Street, Duluth, Minn) reported in this study was developed to accurately determine functional work activities to facilitate a safe return to work after injury or illness. Patients are assessed on their physical performance during generic work tasks. The evaluation emphasizes musculoskeletal safety and function. Maximal functional levels are determined during testing, as well as the physical limiting factors for each task.

To study the relationships of the subjects and their performance, a data bank was established in 1981 by Polinsky Medical Rehabilitation Center. Procedures and scoring for the FCA have remained consistent. All testing therapists were educated during a two-day training session on testing and scoring procedures. In actual testing, these therapists followed a five-hour (over two days) assessment protocol on standardized generic work-related tests.

## PURPOSE

The initial purpose of the data bank was to determine any trends that existed in patient performance. When age and gender consistently appeared to be important variables in the early data study, the tasks more affected by age and gender were selected for particular attention.

To supplement the research on the effects of age and gender, a group of uninjured volunteers also was tested using the complete FCA. This clinical report compares the scores of the uninjured working/active population with performance of the patient population. For the purpose of this study, performance was analyzed on the same tests that had appeared most affected by age and gender in the patient population. These areas are:

1. Weight Capacities
   - Stand up Lift (floor to table height): maximum poundage lifted safely for 10 repetitions.
   - Level Lift (table height to table height): maximum poundage lifted safely for 10 repetitions.
   - Weight Carry (weight carried 100 ft): maximum poundage carried safely and time needed to perform the carry.
2. Stair Climbing
   - Pace: number of stairs per minute.
   - Endurance: maximum time stair climbing could be tolerated, up to four minutes.

3. Balance
   - Forward Walk: time needed to complete 48-ft test and number of errors during performance.
   - Forward Heel to Toe: time needed to complete 48-ft test and number of errors during performance.
   - Backward Heel to Toe: time needed to complete 48-ft test and number of errors during performance.
   - Sideways Walk: time needed to complete 48-ft test and number of errors during performance.

## CONTENT

The data obtained from 1,639 subjects were placed into six categories for computer analysis. For each of the male and female subgroups there were three age categories: young (aged 16 to 34); middle (aged 35 to 55); and advanced (aged 56 to 75). Results are reported by test category.

### Weight Capacities

By examining Table 3A the difference in performance by male and female subjects in different age categories can be determined. In all three weight capacity tests, female subjects had significantly less capacity than their male counterparts.

The change related to age is apparent both in the uninjured population and in the patient population. There is a decline in ability with increasing age. In the uninjured population, the decline in performance for female subjects appeared at an earlier age than did the decline in performance for male counterparts.

Absolute poundages in the patient group are lower than in the uninjured group. In the uninjured group, there are significantly greater abilities in both male and female populations, although the effect of the relationships between age and gender on function is very similar.

### Stair Climbing

The data in Table 3B indicate the difference in performance with age for both male and female subjects in endurance for stair climbing as measured by number of minutes stair climbing could be tolerated. This is particularly evident in the patient population, most of whom were not able to complete

**Table 3A** Comparison of Patient (Injured) Subjects with Uninjured Subjects on Lifting and Carrying Tests

**AVERAGES OF PERFORMANCE**

|  |  | *Injured Female Patients* ||| *Injured Male Patients* |||
| --- | --- | --- | --- | --- | --- | --- | --- |
| Tests |  | Y | M | A | Y | M | A |
| Stand Up Lift (lb) | MN | 23.78 | 19.97 | 16.36 | 39.33 | 33.39 | 28.96 |
| Level Lift (lb) | MN | 34.20 | 30.90 | 21.45 | 48.05 | 43.00 | 37.16 |
| Weight Carry (lb) | MN | 32.15 | 26.99 | 20.54 | 50.87 | 46.43 | 37.23 |

|  |  | *Uninjured Female Subjects* ||| *Uninjured Male Subjects* |||
| --- | --- | --- | --- | --- | --- | --- | --- |
|  |  | Y | M | A | Y | M | A |
| Stand Up Lift (lb) | MN | 41.92 | 28.37 | 14.32 | 67.41 | 64.44 | 51.85 |
| Level Lift (lb) | MN | 59.16 | 49.51 | 48.72 | 82.70 | 82.24 | 73.66 |
| Weight Carry | MN | 61.28 | 51.92 | 52.84 | 84.66 | 84.24 | 77.55 |

*Abbreviations:* A, advanced age category (aged 56 to 75); M, middle age category (aged 35 to 55); MN, mean; Y, young age category (aged 18 to 34).

*Source:* Courtesy of FCA Network, Polinsky Medical Center, 530 East Second Street, Duluth, Minnesota.

the four minutes of the stair climbing test. It is only slightly evident in the uninjured population, as all the young and middle-aged subjects were able to complete the four minutes of stair climb. Those that were not able to, however, were in the advanced age category.

Age in relationship to pace (number of stairs climbed per minute) also is observable. The mean stairs per minute declined noticeably in each increasing age group. Pace declined at a greater rate in the patient population, but the decline of pace in the uninjured is substantial.

Regarding the effect of gender on stair climbing, in the patient population female subjects had less endurance than their male counterparts. This

**Table 3B** Comparison of Patient (Injured) Subjects with Uninjured Subjects on Stair Climb Test

**AVERAGES OF PERFORMANCE**

|  |  | Injured Female Patients ||| Injured Male Patients |||
| --- | --- | --- | --- | --- | --- | --- | --- |
| Tests |  | Y | M | A | Y | M | A |
| Stair Climb—Endurance (min) | MN | 3.49 | 3.16 | 2.97 | 3.70 | 3.41 | 3.28 |
| Stair Climb—Pace (# stairs/min) | MN | 67.69 | 47.77 | 36.08 | 68.09 | 54.90 | 48.64 |

|  |  | Uninjured Female Subjects ||| Uninjured Male Subjects |||
| --- | --- | --- | --- | --- | --- | --- | --- |
|  |  | Y | M | A | Y | M | A |
| Stair Climb—Endurance (min) | MN | 4.00 | 4.00 | 3.86 | 4.00 | 4.00 | 3.95 |
| Stair Climb—Pace (# stairs/min) | MN | 98.79 | 80.22 | 73.45 | 100.00 | 92.16 | 78.96 |

*Abbreviations:* A, advanced age category (aged 56 to 75); M, middle age category (aged 35 to 55); MN, mean; Y, young age category (aged 18 to 34).

*Source:* Courtesy of FCA Network, Polinsky Medical Center, 530 East Second Street, Duluth, Minnesota.

does not appear to be true in the uninjured population; again, because most subjects finished the test. Regarding pace, female subjects had a slower pace than male subjects both in the patient group and in the uninjured group.

### Balance

The data in Table 3C show that age is an important variable both in the number of errors and in the time needed to complete the balance tests. Both in the patient population and in the uninjured population the time needed to complete the balance tests increased with age in all four balance categories. The number of errors increased with age in both the patient and the uninjured populations in all four categories with only one exception. Therefore, those with greater age took more time, which may indicate more effort, and yet made more errors in the balance tests.

*Functional Capacity Evaluation* 189

**Table 3C** Comparison of Patient (Injured) Subjects with Uninjured Subjects on Four Balance Tests

**AVERAGES OF PERFORMANCE**

| Balance Tests | | Injured Female Patients | | | Injured Male Patients | | |
|---|---|---|---|---|---|---|---|
| | | Y | M | A | Y | M | A |
| A: (min) | MN | 0.60 | 0.81 | 1.01 | 0.55 | 0.68 | 0.68 |
| B: (min) | MN | 1.14 | 1.36 | 1.81 | 1.02 | 1.14 | 1.14 |
| C: (min) | MN | 1.64 | 1.88 | 1.78 | 1.30 | 1.44 | 1.40 |
| D: (min) | MN | 0.84 | 1.05 | 1.35 | 0.75 | 0.94 | 1.00 |
| A: (# errors) | MN | 0.09 | 0.48 | 1.57 | 0.12 | 0.30 | 0.53 |
| B: (# errors) | MN | 0.32 | 1.19 | 3.02 | 0.30 | 0.73 | 1.37 |
| C: (# errors) | MN | 1.61 | 2.35 | 4.40 | 1.15 | 1.79 | 2.80 |
| D: (# errors) | MN | 0.27 | 0.42 | 0.97 | 0.25 | 0.39 | 0.56 |

| Balance Tests | | Uninjured Female Subjects | | | Uninjured Male Subjects | | |
|---|---|---|---|---|---|---|---|
| | | Y | M | A | Y | M | A |
| A: (min) | MN | 0.36 | 0.42 | 0.45 | 0.37 | 0.37 | 0.41 |
| B: (min) | MN | 0.81 | 0.89 | 1.09 | 0.73 | 0.78 | 0.82 |
| C: (min) | MN | 1.24 | 1.24 | 1.75 | 0.98 | 1.13 | 1.21 |
| D: (min) | MN | 0.54 | 0.59 | 0.72 | 0.48 | 0.53 | 0.56 |
| A: (# errors) | MN | 0.04 | 0.07 | 0.16 | 0.08 | 0.12 | 0.04 |
| B: (# errors) | MN | 0.20 | 0.22 | 0.88 | 0.13 | 0.48 | 0.59 |
| C: (# errors) | MN | 0.92 | 1.22 | 2.66 | 0.54 | 0.76 | 2.66 |
| D: (# errors) | MN | 0.20 | 0.04 | 0.60 | 0.21 | 0.00 | 0.70 |

*Abbreviations:* A, advanced age category (aged 56 to 75); M, middle age category (aged 35 to 55); MN, mean; Y, young age category (aged 18 to 34).

*Source:* Courtesy of FCA Network, Polinsky Medical Center, 530 East Second Street, Duluth, Minnesota.

Female subjects took longer to complete the balance tests compared to male subjects in all four balance activities and in both the patient and the uninjured populations, with only one exception. Regarding errors, female subjects in the patient population made more errors in the balance tests, but the difference between male and female subjects was less clear in the uninjured subjects.

In balance, then, age is a factor in both errors and time in the patient and in the uninjured subjects. Gender is less clear. Female subjects took longer to complete the tests in both the patient and the uninjured groups, and they made more errors than male subjects only in the patient population.

## PROFESSIONAL IMPLICATIONS AND CLINICAL RELEVANCE

The following three conclusions can be made from examining the data in this study:

1. As a group, the patient population, both male and female, younger and older, scored sharply lower on all test capacities reported. The uninjured population appeared much more functional and capable as an overall group.
2. Female subjects generally had less capacity than male subjects in all weight items. They also appeared to have less endurance and slower pace in stair climbing. In some balance activities they performed less ably than male subjects. This was true both in the patient and in the uninjured populations.
3. Age affects performance on all reported functional activities. Decreased amounts of weight, decreased pace and endurance on stairs, and increased errors and effort on balance all appear a factor of increasing age.

In looking at the affects of age singly and gender singly the older worker and the female worker have lesser capacities and therefore may be at greater risk on the job or at home. The older female worker, by virtue of both age and gender, would be at the greatest disadvantage of any of the six groups studied.

Clinicians can use the information on functional performance to enhance their ability to determine work capabilities of work groups in general, to make recommendations for functional capacities assessment and work hardening, and to understand both healthy and injured working populations.

Specifically, for the occupational medicine clinician, there is often a need to make a return-to-work recommendation or one that would restore work function. The following results of this study may be helpful in those recommendations:

First, female workers and older workers may have increased job risk in weighted activities, antigravity endurance activities (stair climbing), and tasks requiring balance. There may be work activities that involve a combination of these factors, which may place a worker at an even greater risk.

Second, because patients who might return to work are shown to have significantly decreased abilities compared to uninjured subjects, those patient-workers may be at greater risk. Therefore, work hardening programs can be of importance to increase the functional level up to that of the average worker. If a restoration program is not possible, the low patient scores would indicate that physical restrictions would be of increased importance on return to the job. It also points out the necessity for a thorough, individual functional capacities assessment to allow the individual worker to have specific functional return-to-work recommendations.

Third, job analysis is particularly important for the total workplace in identifying when physical lifts and carries, aerobic conditions, and balance are needed on the job. This job analysis might then be compared to the type of work force present at the job and may give indications if age and gender limitations might be present in any particular work condition.

Finally, pre-employment and preplacement screening will be helpful in identifying abilities and limitations on an individual basis.

## RECOMMENDATIONS FOR FUTURE WORK

The three functional areas that were discussed—weight capacities, stair climbing, and balance—would be excellent areas for further specific research. Variables other than age and gender also may be studied to clarify the full relationship on functional abilities. Longitudinal studies also would be helpful to determine whether each person ages in the same manner (as indicated by the means of the different age groups).

Additional external factors that would be helpful to study in relationship to age and gender would be diagnosis, height, weight, exercise levels, and health habits.

Investigations on safety at the workplace by age and gender groups on the functional items reported in this study also would be helpful. This may then correlate apparent functional weaknesses with actual safety at the worksite.

It is hoped that the information gained in this study will provide clinicians with clearer direction for future evaluation and restorative care.

# PART VI

# Work Hardening

Rehabilitation is a multifaceted approach to restore a high level of function to a patient after injury. Different types of injury may require specialized forms of rehabilitation, and the diffentiation of goals will direct the rehabilitative process. Work hardening is the rehabilitation of an injured worker with the goal of attaining a physical level of competence that will allow a return to work. It is implemented after the patient has medically stabilized, physical deficiencies that will interfere with work have been identified, and specific goals have been set for the restoration process.

The aspects of work hardening that set it aside from traditional treatment or rehabilitation are:

1. *Goal specific:* The return to function and activities of daily living may be early objectives, but the successful return to work is the ultimate product of this treatment system.
2. *Job directed:* A functional job description should be used to assist in goal setting. In addition to preparation of physical attributes, there will be an incorporation of work-related activity to facilitate realism in occupational situations.
3. *Program intensive:* A professionally supervised work hardening program maximizes the time during the day and the days of treatment in a week but, concurrently, minimizes the total length of stay in the program. The program intensity both encourages faster progress and decreases the time off work and the physical and compensation costs associated with it.

Benefits of work hardening include increased safety of work functions, increased productivity through increased physical confidence, and an appropriate job match that reduces reinjury.

**Section 15   Work Hardening**

The concept of work hardening in the rehabilitation of the injured worker, and its application to case resolution and maximum rehabilitation, is described.

# Section 15

# Work Hardening

*Carol Franz Lett, Naomi Elizabeth McCabe, Anne Tramposh and Suzanne Tate-Henderson*

## INTRODUCTION

At no other time in the history of rehabilitation has such a multitude of professionals been so involved in the management of the injured worker. Of utmost concern is the physical rehabilitation of the worker who has been identified as being deconditioned and unable to safely manage physical job demands.

As methods of assessing the injured worker become more objective and sophisticated, so must the services to address the problems identified. Accountability, relevancy, cost efficiency, and outcome prediction are desired attributes. Work hardening, serving as the *rehabilitation* aspect of the return to work process, can be adapted very well to these guidelines. A universal body of knowledge related to work hardening is growing.

## NEED FOR WORK HARDENING

The need for work hardening services arises partly from the workers' compensation system. Various "players" in the system include the worker, the employer, the physician, the insurance carrier, and possibly others, such as attorneys and the worker's family.

The traditional medical model tends to foster fragmented services that may result in a case becoming complicated. Traditionally, medical treatment has focused solely on the physical needs of the worker rather than on various other factors. When these other needs are not addressed, the worker can become frustrated, leading to a geometric progression of frustration felt by the other parties as well. Some of these frustrations may be complicated by differences in medical opinions, a "disability mindset" on the part of the patient, social and financial pressures, hostility on the part of the patient and of the employer, and lack of trust between two involved parties.

## WORK HARDENING, AN EVOLUTIONAL PROCESS

The term *work hardening* came into being as professionals used work as a treatment or evaluation modality. Other terms used synonymously are work conditioning, work readiness, work capabilities, and work out–work up.

> Matheson and colleagues define work hardening as: A work-oriented treatment program that has outcome which is measured in terms of improvement of the client's productivity. This is achieved through increased work tolerances, improved work rate, mastery of pain (through the effective use of symptom control techniques), increased confidence and proficiency with work adaptations or assistive devices. Work hardening involves the client in highly structured, simulated work tasks in an environment where expectations for basic worker behaviors (eg, timeliness, attendance and dress) are in keeping with work place standards.
>
> The ultimate goal of Work Hardening is to help the client achieve a level of productivity that is acceptable in the competitive labor market.[1]

This definition underlines the importance of a worklike milieu and expectations of the worker that will effect a positive change toward his or her ability to work. In whatever manner a program is designed to implement the concept of work hardening, the outcome must result in such change as it relates to productivity and the actual workplace.

People frequently enter the rehabilitation process with attitudes of distrust, anger, and a disbelief in their ability to perform, especially in relation to their old job. Through a rehabilitative environment that focuses on special needs of the work-related injury, a transformation stage is set. A concentrated effort will be required for professionals working with injured people to understand the sensitivity of what is to take place.

## LITERATURE REVIEW

A review of the literature in related professional journals reveals that the use of work as a therapeutic and vocational means of rehabilitation is not new. It has threaded in and out of restorative efforts with varying degrees of application and appreciation for a long time. The fields of occupational therapy and vocational rehabilitation have regarded this theme as central to their existence.[1-3]

Since the early part of this century, practical, work-related activities have been used by occupational therapists (1) to improve a patient's strength, endurance, coordination, and range of motion; and (2) to reduce symptom preoccupation through distraction, reinforcement of well behaviors, and development of the capacity to work.[2] Work activities also have been used by occupational therapists and vocational evaluators to assess a patient's ability to work.

Many types of people have benefited from work-oriented therapy and evaluation over the years, including the mentally ill, the congenitally handicapped, the blind, and the elderly; victims of heart attacks, stroke, and chronic disease (eg, tuberculosis); victims of war and war-related injuries; and amputees.[3-6]

The list of diseases evaluated and treated is as long as the list of types of work-oriented programs and approaches. Patients have participated in programs that use crafts and related activities to transfer work skills to actual gainful employment.[3] They also have been placed in actual jobs, whether on light duty or under heavy supervision. Sheltered workshops also have been used, as well as work samples and actual job simulation.[5,7-9]

Gradually, terminology and methods have evolved to the point where they more specifically address the issues surrounding worker-related injuries. The most recent systems of work evaluation assess performance relating to a particular job or a patient's physical capabilities as they relate to work in general. Most systems of work evaluation have the major purpose of "determining needs, measuring abilities and predicting capacities of an individual."[9] They are "yardsticks by which a tester can determine capacity for a variety of physical tasks."[10] Some examples of capacity measures are lifting, kneeling, pushing, pulling, manipulating, grasping, writing, and perceptual and cognitive skills as they relate to physical performance. Also assessed is a patient's so called "feasibility for employment."[11-14] Characteristics, such as punctuality, attention to a task, and attendance, are observed and recorded.

From the patient's standpoint, there also have been benefits in work-oriented services. One such benefit is the prevention of iatrogenic illness, which occurs with the past and current medical model. The scope of the iatrogenic illness problem is cited frequently in the literature. Nachemson[15] recognizes that if patients with low back pain were properly treated, many of them would be free of symptoms more quickly and some of the severe effects of long-term confinement and inactivity would be avoided.

Many nonmedical reasons are cited for lack of successful return-to-work outcomes. Wilke and Sheldahl report these reasons include "unwarranted medical restriction, patient anxiety concerning ability to meet job demands, apprehensive family members and fearful employers."[6]

Work-related therapeutic intervention has been shown to reduce the lack of return-to-work outcomes by increasing physical endurance, decreasing symptoms and increasing patient self-confidence.[9] Well behavior is reinforced by normalizing the patient's routine and shifting responsibility for physical well-being from the medical community to the patient.[16]

Work-related therapy also makes the patient "face the issues" in a timely manner and discover whether or not he or she is physically capable of performing the former job.[17] If the patient is not able to return to the old job, facing the issue will permit action toward seeking other employment.

## TYPES OF WORK HARDENING SERVICES

A variety of program types seem to be most evident at this point in the evolution of work hardening. Their needs regarding space, equipment, staff, and methods differ.

### Physical Therapy Model

Programs staffed solely by physical therapists tend to be in private clinics in a hospital as an extension of the physical therapy department. Often the focus of these programs is primarily on the physical aspects of work-related performance or on physical conditioning—ie, to improve the person's physical conditioning so that the potential for work is increased.

Frequently, a work hardening program is a later component of primary physical therapy begun during the immediate postinjury phase of recovery. Certain primary care modalities, including hot and cold packs, massage, ultrasound, and electrical stimulation, continue to be used throughout the exercise-oriented program. If a patient has met the established goals, it is expected that self-confidence will come from the increase in mobility and strength and this will prepare the worker for the challenge of work.

Goals of upgraded strength and endurance might be combined with forms of pain management, such as biofeedback, the Transcutaneous Electronic Nerve Stimulator (TENS), and lectures and demonstrations on good body mechanics. Specialized body building equipment, such as Nautilus, Nordic Track, Mini-Gym and Cybex, may be used. Other traditional equipment might include barbells, weights, treadmills, and exercycles.

Depending on space, common materials, including crates, bricks, concrete blocks, wheelbarrows, handtrucks, and shelves, also may be used.

Fitness clubs are often used to provide physical space for the physical therapy or to provide additional time and exercise for the patient program. Usually, graded home exercises and supportive instructional materials are provided to reinforce what has been accomplished. On-site job analysis may or may not be a component of this style of work hardening program. It is frequently difficult for a therapist to leave a center, especially if there is a mixed caseload, which includes people who are not there for work hardening. Therapists who provide back schools to industry, however, are beginning to incorporate information and experience gleaned from work hardening and give more balance to the content of such schools (eg, ergonomic analysis of the work station and adaptive tool recommendations).

The physical therapy model is extremely useful in uncomplicated cases when the major barrier to return to work is physical deconditioning secondary to illness and/or injury. It also is important when instruction in correct body mechanics, posture, and a general understanding of the injury or illness are needed. This type of program is especially suited to the highly motivated person interested in an expedient, aggressive rehabilitation program.

The patient may begin the work hardening phase of the program with 30 minutes of pure stretching and strengthening exercises and then progress by increasing exercise time and tolerance. Gradually more activities are added, (eg, swimming, Nautilus circuits, exercycle) until the client is involved in the program three to four hours per day, three to five times per week.

The client is discharged when maximum physical recovery has been reached and it is believed that he or she can return to the job. A functional capacity evaluation is often chosen to provide return-to-work information.

*Case Study*

Mr. G. is a 40-year-old man who was injured on the job as a truck driver. The injury occurred when he slipped on ice while getting into his truck. He sustained a strain to his lower back. Treatment included bed rest and medications for approximately 8 weeks. Mr. G. attempted to return to his job but was unsuccessful. He stated he was unable to sit and had difficulty lifting more than 20 to 30 lb. He was referred by an orthopedic physician for physical therapy evaluation and treatment.

Mr. G. was evaluated, and it was determined that his primary problems were the following:

1. Decreased lower extremity strength and endurance.
2. Poor body mechanics.
3. Poor sitting posture.
4. Subjective complaints of pain with forward flexion and prolonged sitting.
5. Decreased general condition.

Mr. G. began his program with individualized and group instruction in body mechanics, posture, and anatomy. He also began a general conditioning program consisting of bicycling, lower extremity Nautilus circuit training, general stretching exercises, and swimming. He was scheduled initially three times per week, one hour per day and increased gradually to four hours per day, three days per week. Mr. G. was well motivated and showed excellent progress in all problem areas. After 4 weeks Mr. G. was discharged and expressed confidence that he could handle the requirements of his job. He complained of pain only on arising in the morning. He was instructed in a home program before discharge. Six months' follow-up indicates Mr. G. is doing well.

**Occupational Therapy Model**

Most occupational therapy programs emanate from larger facilities (eg, rehabilitation clinics, hospitals, or comprehensive hand therapy units). Emphasis on return to work functions is a natural extension of the holistic approach of occupational therapy in promoting a person's independence in activities of daily living.

Components of an occupational therapy program generally reflect an activity-oriented approach, allowing the client to rehearse the actual activity patterns and sequences required for work performance. Equipment includes fabricated work stations, which are designed to stress body positions or motor patterns, and manual or power tools, which are used to build tangible projects. Another example is a sophisticated commercially available "work simulator." All equipment, however, allows for grading of difficulty in terms of complexity, time, and resistance involved. Because of the strong functional emphasis of this program, on-site inspection of the job may be a program service to allow for more job specific practice and conditioning sessions.

Therapists also may use certain assessment tools that provide information regarding mental/emotional status and recorded observation of the client's work habits, peer interaction, authority compliance, direction comprehension, and a host of other behaviors affecting work performance.

The program is generally housed within the occupational therapy department or rehabilitation clinic, possibly with a separate space designated for work hardening. Nevertheless, this may provide for a smooth transition from the more acute outpatient treatment to work hardening.

*Case Study*

L. B. is a 35-year-old dock worker for a freight company. After non-union of a left scaphoid fracture he underwent surgery for a tri-scaphoid fusion. After removal of the cast and pins, work rehabilitation efforts were initiated 1 month postoperatively in an outpatient hand rehabilitation clinic. The client was seen for 1 to 1½ hours a day during the first week of treatment for range-of-motion exercises to the wrist and digits, mild resistive putty exercises for strengthening, and dexterity exercises in elevation for edema reduction.

The following week a work tolerance screening and job analysis was performed by the therapist, which revealed that the client was able to lift and carry 35 lb safely in comparison to the 75-lb job requirement. His work tolerances were limited primarily by decreased left hand and wrist strength and endurance, and painful and decreased left wrist range of motion.

He was started immediately on a work hardening program consisting of nonresistive reaching and handling tasks for increasing range of motion, functional repetitive strengthening exercises on the work simulator, repetitive lifting, carrying and stacking of 20 to 35-lb objects, and a woodworking project requiring bilateral use of both manual and power tools. He participated in work hardening 4 hours a day for 2 weeks with a gradual increase in weights and resistance until (during the last 3 days) he was performing lifting, carrying, and stacking tasks for a period of 2½ hours with various sized crates and boxes ranging from 25 to 80 lb. His timeliness, work pace, attitude, and adherence to safety precautions were all closely observed and documented. Discharge from the program with return-to-work release was recommended following 3 weeks of outpatient occupational therapy care.

## Occupational Therapy and Physical Therapy Model

When occupational therapy and physical therapy services are available within the same outpatient center or department the two separate models can easily be combined. Each discipline and its approach in the work hardening process can be the complement of the other and actually provide the client with greater assurance of total conditioning and preparedness for return to work.

The physical therapy and occupational therapy team approach is widely accepted in the medical community and assures provision of services addressing the physical, functional, and behavioral aspects of the client.

*Case Study*

Mr. P. is a 38-year-old factory worker for a bottling company. He sustained a lumbar strain while stacking cases of empty bottles onto a conveyor belt. This was his third low back injury on the same job and it was questioned whether Mr. P. was suited for this type of work.

Mr. P. was treated with traditional physical therapy modalities and exercise and was tested for his physical capacities by the occupational therapist. Significant physical findings compared to job requirements were the following:

| Physical Capacities | Job Requirements |
| --- | --- |
| 1. Lifting 0 to 45 in, 45 lb | Lifting 0 to 45 in, 50 lb |
| 2. Repetitive squatting, unloaded 12 repetitions per minute—four minutes | Occasional repetitive squatting required |
| 3. Sustained squatting/kneeling 15-minute duration | Occasional sustained squatting/kneeling 30-minute duration |

The major limiting factors were decreased general condition and extremely unsafe body mechanics.

Because of the previous history of re-injury, it was believed that Mr. P. could benefit from work hardening primarily to practice proper techniques to prevent re-injury.

Mr. P. was instructed in body mechanics and job simulation was set up by the occupational therapist. Mr. P. also began a general conditioning program consisting of strengthening, stretching exercises, and aerobic conditioning with treadmill and bicycle. He was seen on a daily basis for three hours a day.

After 3 weeks, Mr. P. demonstrated the ability to perform safely the simulated task for a duration of two hours. He felt confident that he would be able to perform his job safely on discharge.

## Vocational Rehabilitation Counselor Model

Historically, "work adjustment programs" have been developed primarily by vocational rehabilitation counselors. Primarily, these programs have been used when it is not feasible for a person to work because of disease or injury. The goal of the program is eventual employment even if restricted to a sheltered situation. Many of the concepts are the same as those in the occupational therapy model in that work is used as a medium for rehabilitation. Some programs also use commercially available work samples for testing and task simulation. Compared with the other models presented many of these types of programs involve extended time frames.

The vocational rehabilitation counselor also may be involved with a work hardening program as a department team member or be a referral source through a private rehabilitation management company.

Work hardening simplifies the task of vocational rehabilitation by raising the work capacity of an injured worker to a level sufficient to permit his or her return to the previous job. The rehabilitation counselor may be needed to clarify certain aspects of that return for the worker or to bolster and support the confidence to do so.

It is more desirable and appropriate for the worker to return to his or her old job from a cost and time frame standpoint. The exception to this would be a situation in which the job is unsafe and causes acute or cumulative trauma. The solution, then, is not to upgrade the worker but to reevaluate the ergonomics of the work station. Another exception would be a situation in which the worker could not physically perform the requirements of the job.

The issue of return to work becomes more complex if the person cannot return to the old job and must be considered for a new job requiring various degrees of new skills or education. In this situation the role of the rehabilitation counselor takes on added meaning. The rehabilitation counselor, then, may administer various intelligence, aptitude, and interest tests to determine worker traits that can be compared to job requirements. Examples of such tests are the Slosson Intelligence Test, the Wide Range Achievement Test and the Vocational Preference Inventory. Functional tests, such as the Valpar and Singer component work samples, also may be administered to obtain an objective sampling of the client's work skills.

The rehabilitation counselor also may counsel the client on job-seeking skills or on types of jobs available within his or her capabilities. Counselors may or may not be involved in actual job placement, depending on whether the facility in which the counselor works provides this service.

*Case Study*

J.B. was referred to a vocational rehabilitation counselor while participating in an occupational-physical therapy work hardening program. Based on the physical capacities testing performed by the occupational and physical therapists, Mr. B. was functioning at a sedentary-light level. His former job as a hod carrier for a construction company was classified as very heavy. Mr. B. had undergone a lumbar laminectomy and fusion with Harrington rods 1 year before referral. It was highly unlikely that Mr. B. would be able to return to his former occupation.

The vocational rehabilitation counselor obtained a complete work history designed to determine transferable skills. A battery of tests were then administered to determine Mr. B.'s interests, personality characteristics, intelligence, and verbal and communication skills. The counselor also worked with Mr. B. on job-seeking skills. Mr. B. had expressed interest and had shown aptitude for the real estate field and had actually begun exploring his options along these lines. With the assistance of the vocational rehabilitation counselor he further pursued this avenue, completed his license requirements, and obtained a position as a real estate agent. On 6 months' follow-up this client actually expressed gratitude that his injury had occurred and forced him to pursue another career.

## Comprehensive Psychophysical Model

The comprehensive psychophysical model is based on a multidisciplinary approach. Many services in hospitals or outpatient rehabilitation centers are using this approach.[16,18]

The rationale for using a holistic team approach is well accepted. It has been used successfully with many different treatment protocols, including the treatment of chronic pain in a pain management program, the treatment of head trauma and spinal cord injury, and, now, the treatment of work-related injuries.

Over the years, the multidisciplinary approach has gained popularity in many rehabilitation settings because of the complex nature of so many injuries and illnesses. It has long been recognized that the negative interaction among a patient's physical, social, and psychological functions can complicate disease. It is well-established that in comparison with a single rehabilitation approach the multidisciplinary approach has provided more successful outcomes in dealing with the complex injuries and illnesses that predominate.[5,16,19-21]

In a work capacity service rehabilitation center, team members might include physical therapists, occupational therapists, rehabilitation counselors, industrial science specialists, psychologists, social service workers, a business office coordinator, and a center manager.

If a team cannot afford the services of a permanent psychologist or the larger facility requires that professional's attention to other programs as well, it is possible to gain this type of input through the consultation services of an industrial psychologist who has a particular aptitude for applying evaluation results to industrial situations. This person can also provide professional support to the staff members for their own needs.

The comprehensive psychophysical model is an excellent example of how all the needs that can arise within a given case can be understood while the right resources to deal with these needs in the most appropriate way can be provided.

In this model, each team member functions as a work capacity specialist. This title promotes the purpose of the center's existence to the worker, the physician, the employer, and the carrier. Every team member is highly experienced in his or her own field but also shares certain cross-training responsibilities with the other members. For example, a physical therapist may primarily cover the exercise classes but an occupational therapist may back up this person by doing the same activities when the work demands call for it. Likewise, physical therapists learn the components of the various work stations that the occupational therapist and industrial science specialist usually oversee. This arrangement permits flexibility in scheduling, procedural reliability, and better understanding among the various professionals regarding each other's work. These advantages are reflected in the quality and consistency of the important reports generated from work hardening services.

With the multidisciplinary team effort, issues surrounding functional impairment and methods for improvement are identified and resolved more quickly and efficiently. This is facilitated by the fact that each staff member works with work-related injuries and no other diagnoses are seen within the center.

When a case is referred to the center, a specific format is followed. The format selected will depend on the circumstances of the case and it must meet the approval of the referral source. All services provided lead to successful case resolution. Special attention is given to the referral source's questions, concerns, and goals for each person. This aids in the determination of the most appropriate services for the individual client.

The comprehensive model for work assessment and rehabilitation has proved to be timely and cost effective; it avoids secondary gains and leads to maximum recovery and return to work. The work hardening process addresses the interaction of both physiological and psychological factors resulting from an injury and incorporates crucial components of medical and vocational rehabilitation.

## AN EXAMPLE PROGRAM

As an example of a work hardening program the following describes the application of one of the types of work hardening services—the comprehensive psychophysical model—in a specific setting. The program was developed by $W_x$: Work Capacities Inc, Shawnee Mission, Kansas and is called Functionally Fit for Work, which reflects the philosophical approach of the center. The program engages the client in a continuum of individual functional work-related services, beginning with the base line evaluation and concluding with the client's physical-emotional readiness for work.

### Initial Evaluation

Before a client is admitted to the program, his or her perceived potential to improve through such a program is thoroughly discussed with the referral source. On admission, the evaluation battery is scheduled. The components of this battery are the following:

1. Intake Questionnaire and Psychosocial Assessment
2. Vocational Assessment
3. Structural (Musculoskeletal) Evaluation
4. Subjective Functional Profile
5. Vocational and Ergonomic Analysis
6. Base Line Functional Analysis

Through evaluation, a person's major limiting factors as they relate to his or her employability are identified. Realistic rehabilitation goals are established by the team and provide an ongoing measurement for progress and achievement.

*Intake Questionnaire and Psychosocial Assessment*

The intake questionnaire is completed before any work is done. The initial interview process is designed to assess a client's psychosocial stature. Short, standardized tests may be administered at this point to assist the team in identification of personal stressors, habits, interests, barriers, conflicts, and level of motivation in the process of reentry to a productive life style.

*Vocational Assessment*

Through the vocational assessment relevant information is gathered from the client's vocational history, the specific work requirements of the job held at the time of the injury, and the client's education and training background and then vocational alternatives are identified. A verbal job analysis may be supplemented later by an actual on-site analysis if further clarification or documentation of physical job demands is needed. In the vocational evaluation, the important factors are what has the client done in the past, what work was being done at the time of the accident, and what work might be able to be done in the future. Vocational aptitude and psychological testing may be used if necessary.

*Structural (Musculoskeletal) Evaluation*

The structural (musculoskeletal) evaluation is a thorough assessment that is essential in the documentation of current objective physical findings, such as existing range of motion, strength, and postures. In most settings, the physical therapist and/or occupational therapist will perform this evaluation and report the results to the other team members who will perform the base line functional analysis.

*Subjective Functional Profile*

The subjective functional profile is the client's report of his or her current level of function in relation to activities of daily living. This is necessary in providing the rehabilitation team with insight into behaviors of self-limitation and an overall picture of the client's activity level and, thus, general physical condition and endurance.

From this information, correlations may be made and consistencies addressed between structural deficits and subjectively reported functional deficits. A client's perception of his or her pain in terms of location, type, and degree can be documented through the use of body diagrams and a numerical scale. Consistencies and/or inconsistencies may be noted later through comparison of subjective reports of pain and functional tolerances and objective findings identified in the structural (musculoskeletal) evaluation and the base line functional analysis.

*Vocational and Ergonomic Analysis*

When an on-site analysis is appropriate, a vocational and ergonomic analysis of the work station is done. This includes objective, reproducible, and factual measurements of the static forces required for job tasks, such as lifting or pulling. Software programs that analyze static strength, body positions, and metabolic energy expenditure are used to arrive at a judgment regarding functional performance of the worker compared to the job requirements.

*Base Line Functional Analysis*

The functions of the base line functional analysis are twofold. First, it provides documentation of a client's physical tolerances as related to work, thus establishing the criteria for the level at which the client enters the work hardening program. Second, it provides base line information for documentation of change when a comparative functional analysis is performed on program completion.

Information obtained from the medical history, psychosocial assessment, vocational information, subjective functional profile, and the structural (musculoskeletal) evaluation is used to formulate the evaluation plan for the base line functional analysis. Evaluation tasks are prioritized according to their physical demands, their frequency of performance on the job, the client's reported limitations in functional tolerances, and the objective findings from the structural (musculoskeletal) evaluation.

The following case study demonstrates the formation of base line functional analysis tasks with consideration of the client's objective and subjective findings.

*Client A*
*Diagnosis:* Two months' status postoperative partial hemilaminectomy L5-S1; excision nucleus pulposis
*Job Title:* Laborer
*Job Duties:* Responsible for laying new track and installing switches on railroad

*Critical Job Demands:*
1. Standing/walking: 10 hours total duration.
2. Lifting: (a) 90-lb jack hammer; (b) one end of 260-lb railroad tie, ground to waist level.
3. Carries 90-lb jack hammer approximately 40 ft, two times per day.
4. Push/pull 800-lb track preliner with both upper extremities, 20 to 30 ft.
5. Stooping: setting rail anchors, sustained and repetitive.
6. Squatting: setting and wrench tightening gauge rods.

7. Reaching: predominantly below knee level; may swing sledge overhead.
8. Manipulating: grasping/holding tools.

*Objective Structural (Musculoskeletal) Limitations:*
1. Decreased mobility in lumbar and thoracic spine, primarily in extension (25% to 50% of normal).
2. Peripheralization of symptoms with flexion in supine position.

*Subjective Functional Limitations:*
1. Sitting, standing, or driving for longer than 30 minutes at a time.
2. Walking for longer than one hour.
3. Lifting/carrying more than 10 to 15 lb
4. Kneeling, squatting for one to two minutes.

*Base Line Functional Analysis Tasks:*
1. Total body range of motion/manipulation
2. Standing/walking
3. Static/dynamic lifting
4. Carrying
5. Static/dynamic push/pull
6. Squatting
7. Stooping

If possible, standardized evaluation equipment should be used in the base line functional analysis to allow comparisons between the job requirements and the client's current level of performance. Explicit documentation should be made regarding repetitions/frequency, duration, distance/range, force/torque, and weight/load. Both subjective and objective observations should be recorded as comparative base line information as set forth in the following examples:

| | |
|---|---|
| *Functional Activity:* | Carrying |
| *Evaluation Task:* | Bilateral carry—12 x 12 x 12 in crate, waist level, elbow flexion at 90 degrees. |
| *Observations:* | |
| *Subjective:* (on a scale of 1 to 10) | Begin pain "2"; end pain "3," more central low back. |
| *Objective:* | 20 lb—100 ft in 37 seconds |
| | 30 lb—100 ft in 37 seconds |

40 lb—100 ft in 36 seconds
45 lb—100 ft in 36 seconds
Reciprocal gait pattern.
Increased trunk extension with 40 to 45 lb.
Slight limp noted when carrying 45 lb.

**Identification of Work-Related Rehabilitation Goals**

The comprehensive evaluation should identify a client's abilities, as well as functional limitations as they relate to work. Physical and psychosocial limiting factors are documented and provide the client and the rehabilitation team with specific areas of weakness toward which to direct work-related rehabilitation goals. The client must be considered a member of the team in the establishment of these goals. It is important that he or she understand current capabilities and limiting factors as they relate to return to work so that rehabilitation goals can be realistic and attainable. Ultimately, it is the client who must take the necessary steps toward achieving these goals and who must be able to detect positive outcomes in the process.

The goal for every person participating in a comprehensive evaluation program is to become employable. Whether this means returning to the same job, a modified job, or a new job depends on the type and extent of injury, the resulting structural and functional limitations, the client's job at the time of injury, and the client's level of motivation and existing barriers in the return-to-work process.

No matter what other goals are identified a client's plan for vocational reentry must be identified from the onset. The team must assist the client to realize that although a complete (100%) return to normal life style is rarely achieved in a 4- to 6-week rehabilitation program, employability is generally attainable if the client's goals are realistic. With this in mind, necessary improvements and short-term goals, as related to return to work, can then be formulated more easily. With specific job demands in mind, the team, including the client, can then establish specific goals in the areas of flexibility and strengthening exercises, posture and body mechanics training, pain management, and work conditioning. Goals in each of these areas must be directly related to the goals for return to work, whether it be to return to the same job, a modified job, or a new job.

Work-related rehabilitation goals must be reassessed continually and adjustments made as necessary to ensure the goals remain realistic and attainable with the client's potential for success in mind. An example of how goals are established in relation to a client's major limiting factors follows.

*Work-Related Rehabilitation Goals*
A. Client's Major Limiting Factors:
   1. Peripheralization of symptoms with lumbar flexion activities and movements, which limits stooping, squatting, sitting, and lifting.
   2. Decreased lumbar mobility, especially extension, which limits standing, reaching, and carrying.
   3. Decreased lower extremity strength and endurance, which limits squatting, lifting, and climbing.
   4. Subjective complaint of lumbar, right lower extremity pain with increased activity.

B. Long-Term Goals at Discharge:
   1. Return to previous job as Order Selector/Forklift Operator.
   2. Be able to lift, carry 50 to 60 lb safely.

C. Short-Term Goals:
   1. Stress/pain management
      (a) Instruct in positions/movements to centralize/stabilize symptoms.
      (b) Lumbar flexibility exercise, primarily extension.
      (c) Complete stress/pain lectures and relaxation training.
   2. Exercise program
      (a) Lumbar flexibility exercises.
      (b) Isokinetic lower extremity strengthening exercises.
      (c) Aerobic conditioning on stationary bicycle, and walking program.

Goal establishment continues as above for each component of the program until all major limiting factors are addressed and a plan for alleviating such limitations is identified.

## Program Philosophies

### The Balance Concept

A comprehensive rehabilitation program, designed to facilitate an injured worker's employability, must encompass five major areas. All components are essential in maximizing a client's productivity. They include:

212  Work Injury

**Figure 15-1**  Flexibility and strengthening exercises. *Source:* Courtesy of $W_x$: Work Capacities Inc, The Center for Work Assessment & Rehabilitation, Westwood, Kansas.

- flexibility and strengthening exercises (Figure 15-1)
- stress and pain management
- posture and body mechanics training (Figure 15-2)
- work conditioning with job simulation (Figure 15-3)
- vocational reentry counseling (Figure 15-4)

A successful approach to return to work is dependent on each interrelated component.

*Daily Work Schedules*

To closely simulate the demands on a worker, a comprehensive program should be designed to require daily participation (Monday through Friday), six to seven hours a day. This allows for a well-rounded and balanced program that focuses on client educational sessions during the early weeks and on assimilation and increased work confidence through practice and

**Figure 15-2** Posture and body mechanics training. *Source:* Courtesy of W$_x$: Work Capacities Inc, The Center for Work Assessment & Rehabilitation, Westwood, Kansas.

---

application of newly learned methods in stress and pain management, posture, and body mechanics during later weeks. As increased physical demands are placed on the client less time is spent on learning and more time is spent on practicing.

The following is an example of a client's schedule in the first week and in the fourth week. There is a gradual increase in productivity while effective pain management techniques, correct postures, correct body mechanics, and adapted methods of job performance are applied.

*Week One*
  8:30 -  9:00   Warm-up exercises
  9:00 -  9:30   Work station activities
  9:30 -  9:40   Break
  9:40 - 10:30   Back education
 10:30 - 11:00   Aerobic exercise
 11:00 - 12:00   Work station activities

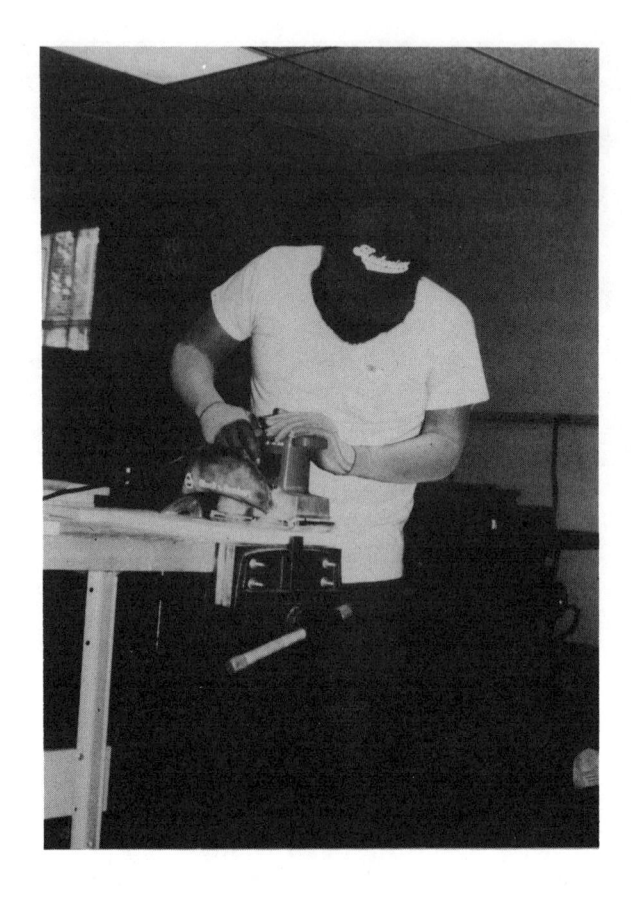

**Figure 15-3** Work conditioning with job simulation. *Source:* Courtesy of $W_x$: Work Capacities Inc, The Center for Work Assessment & Rehabilitation, Westwood, Kansas.

```
12:00 -  1:00   Lunch
 1:00 -  2:00   Work station activities
 2:00 -  2:30   Stress/pain management
 2:30 -  3:30   Woodworking (Special projects)
```

*Week Four*
```
 8:30 -  9:00   Warm-up exercises
 9:00 -  9:30   Work station activities
 9:30 -  9:40   Break
 9:40 - 10:30   Work station activities
```

# Work Hardening    215

**Figure 15-4**  Vocational reentry counseling. *Source:* Courtesy of W$_x$: Work Capacities Inc, The Center for Work Assessment & Rehabilitation, Westwood, Kansas.

---

| | |
|---|---|
| 10:30 - 11:00 | Aerobic exercise |
| 11:00 - 12:30 | Job simulation |
| 12:30 -  1:00 | Lunch |
|  1:00 -  2:00 | Work station activities |
|  2:00 -  2:10 | Break |
|  2:10 -  3:00 | Job simulation |
|  3:00 -  3:30 | Woodworking (special projects) |

*Supportive Group Atmosphere*

A group atmosphere, created by 10 or more clients (ideally) working together simultaneously, not only simulates more closely most work environments but also provides clients with a built-in support system and the comfort of knowing each is not alone in the often frustrating system of workers' compensation claims and processes. The rehabilitation counselor can further encourage the support and sense of camaraderie through group and individual counseling sessions. Specific frustrations, as well as emotional barriers to return to work, can be addressed.

The counselor may conduct psychovocational testing, vocational exploration counseling, and job-seeking skills training, when indicated, in relation to the work-related rehabilitation goals.

*Client Self-Responsibility*

A client's education in and understanding of the following areas is critical to his or her self-responsibility and, in turn, to the attainment of a productive life style:

1. Basic physioanatomy as related to the injury.
2. Prevention and pain management through correct postures and body mechanics.
3. Positive effects of exercise on improving flexibility, strength, and endurance.
4. Methods of self-treatment in the management of stress and pain.

Only when a client understands the above can he or she effectively integrate and consistently apply the methods during functional work performance.

*Challenge Position Work Stations*

A comprehensive work hardening program must provide a variety of work stations that challenge the client not only in work functional positions and movements, but also in attaining work-related goals. (See Figures 15-5 through 15-10.) For example, a construction electrician could logically progress from a job involving fabricating a large floor-based wooden structure and requiring standing, reaching floor to overhead and using of small tools to a job simulation task requiring intermittent climbing, sustained reaching, and actually performing electrical wiring. Whereas the first work station allows for intermittent rest breaks for pacing purposes owing to its "start and stop" design, the second requires sufficient endurance to complete the "job."

*Job Simulation*

Progress in work conditioning is made in relation to duration, frequency, and load, with specific work-related rehabilitation goals in mind. The team should be prepared to adapt current equipment or fabricate new equipment to simulate as closely as possible the components of a client's job. The client generally engages in specific job simulation tasks for 1 to 2 weeks before

Work Hardening 217

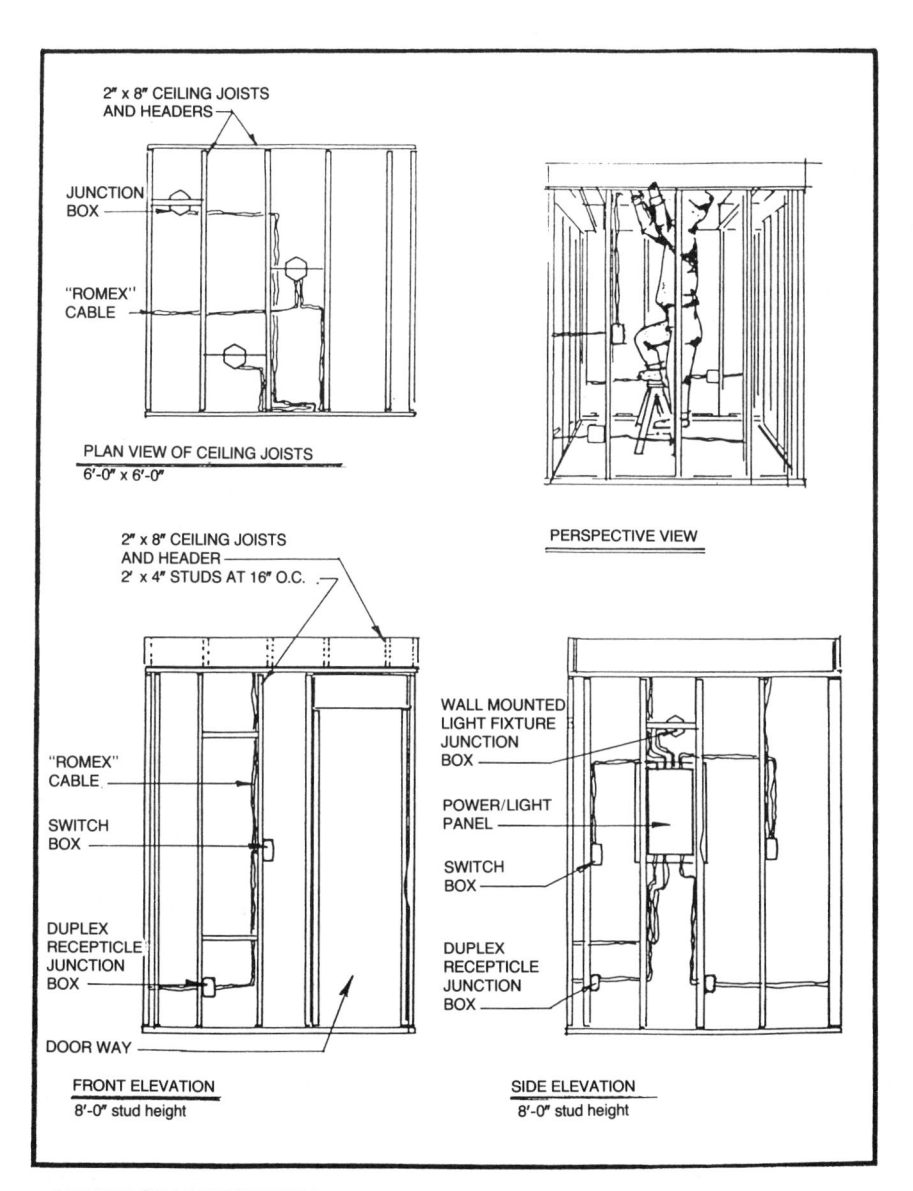

**Figure 15-5** Construction electrician work simulation. *Source:* Reprinted from *Work Capacities' Technical Training Manual* (p. 381) with permission of Work Capacities Inc. © 1986.

218  Work Injury

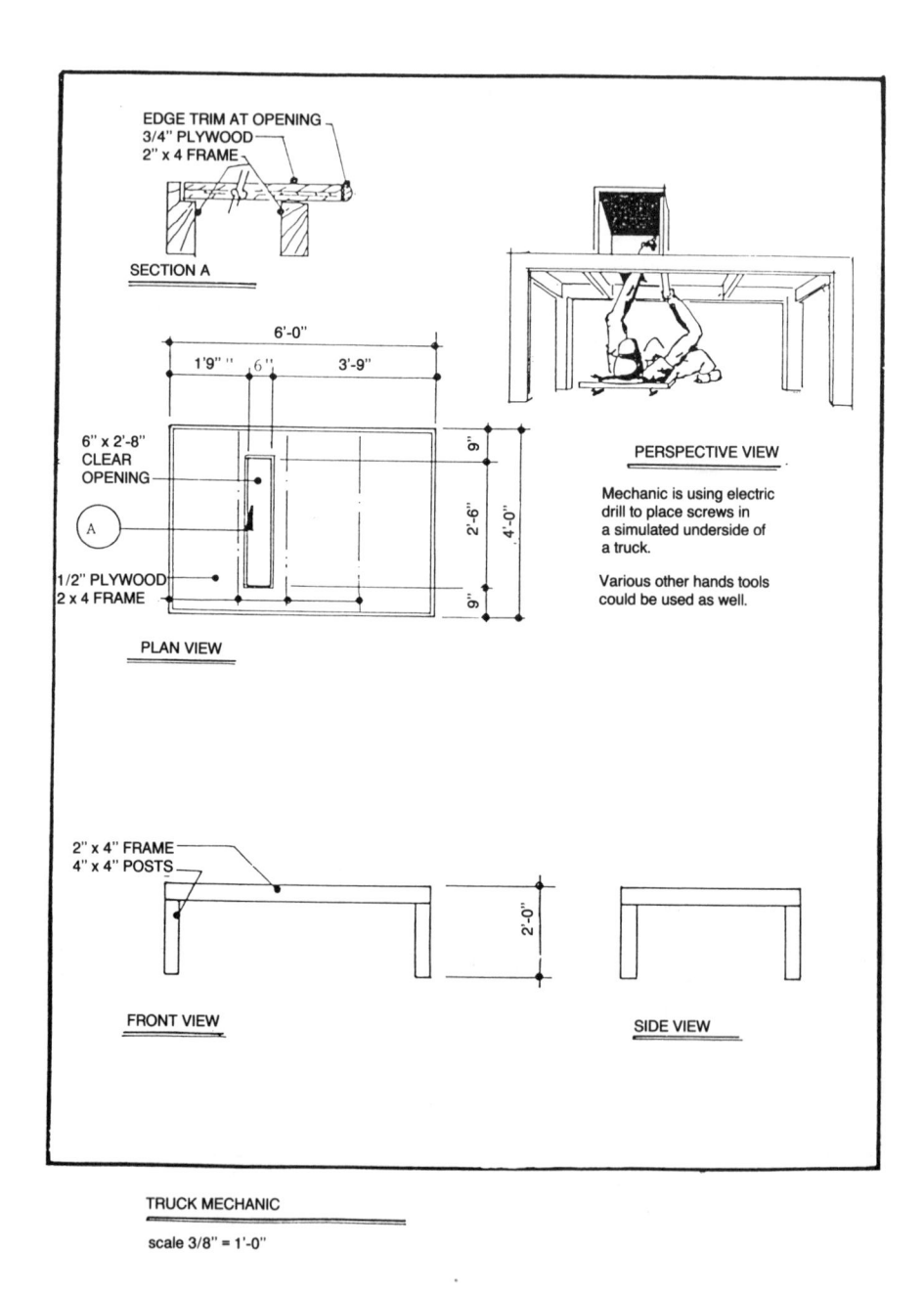

**Figure 15-6**  Truck mechanic work simulation. *Source:* Reprinted from *Work Capacities' Technical Training Manual* (p. 383) with permission of Work Capacities Inc. © 1986.

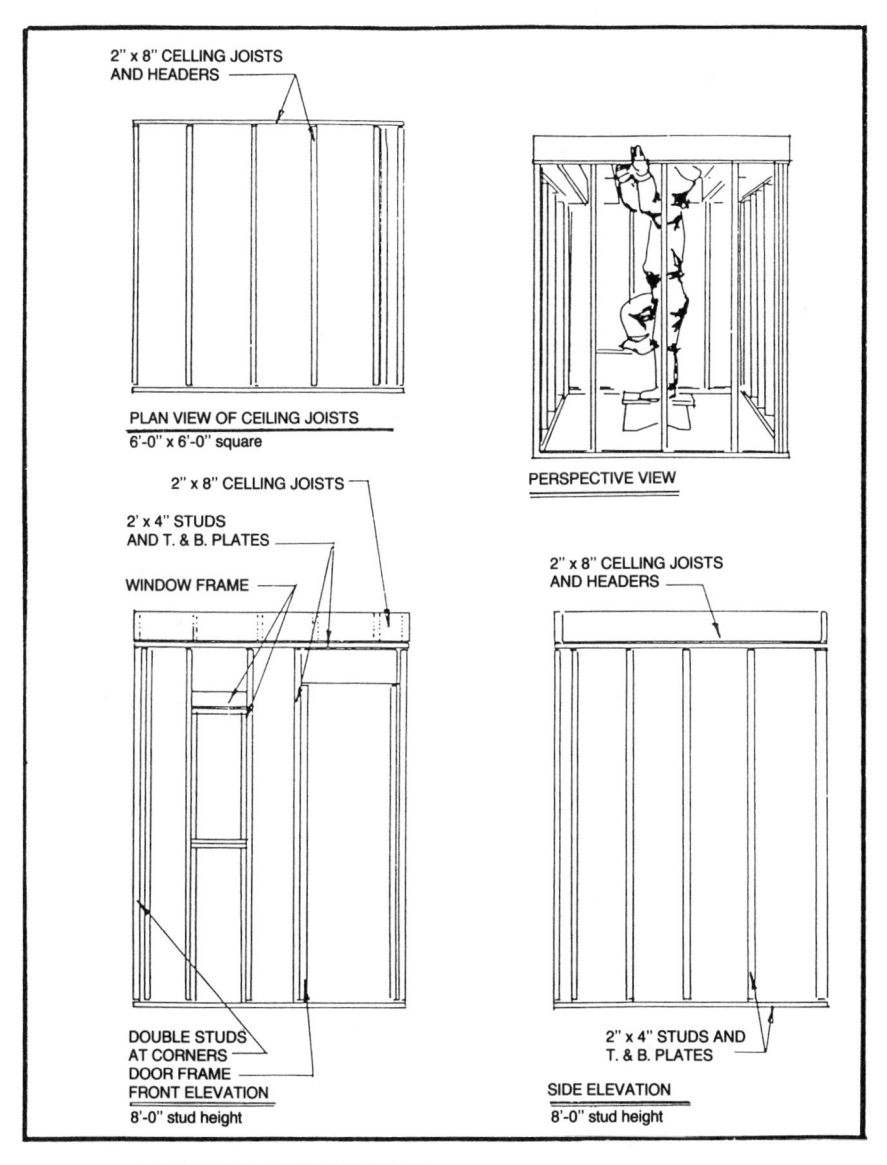

**Figure 15-7** Construction framing carpenter work simulation. *Source:* Reprinted from *Work Capacities' Technical Training Manual* (p. 385) with permission of Work Capacities Inc. © 1986.

220  WORK INJURY

**Figure 15-8** Assembly boards work simulation. *Source:* Reprinted from *Work Capacities' Technical Training Manual* (p. 386) with permission of Work Capacities Inc. © 1986.

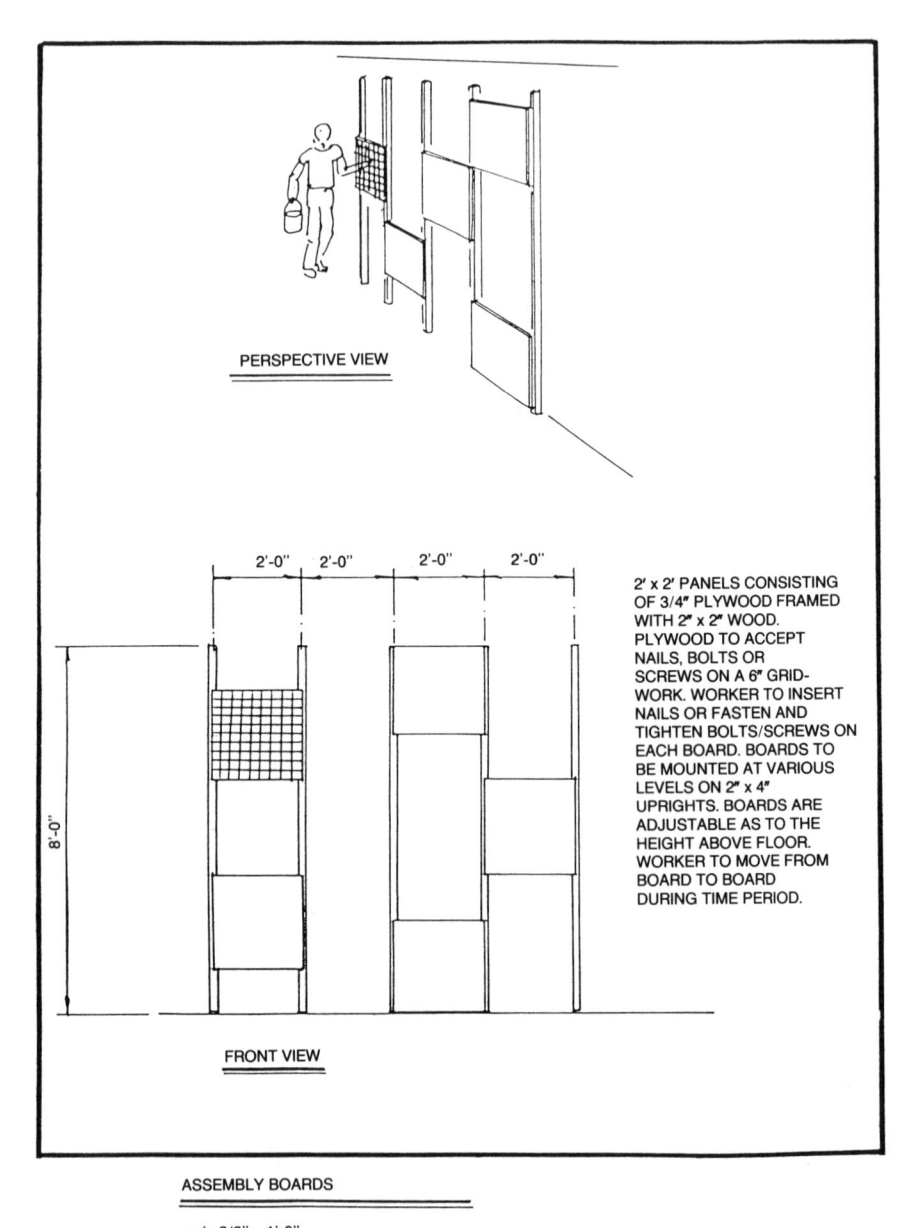

Work Hardening 221

**Figure 15-9** Ramp and bins work simulation. *Source:* Reprinted from *Work Capacities' Technical Training Manual* (p. 384) with permission of Work Capacities Inc. © 1986.

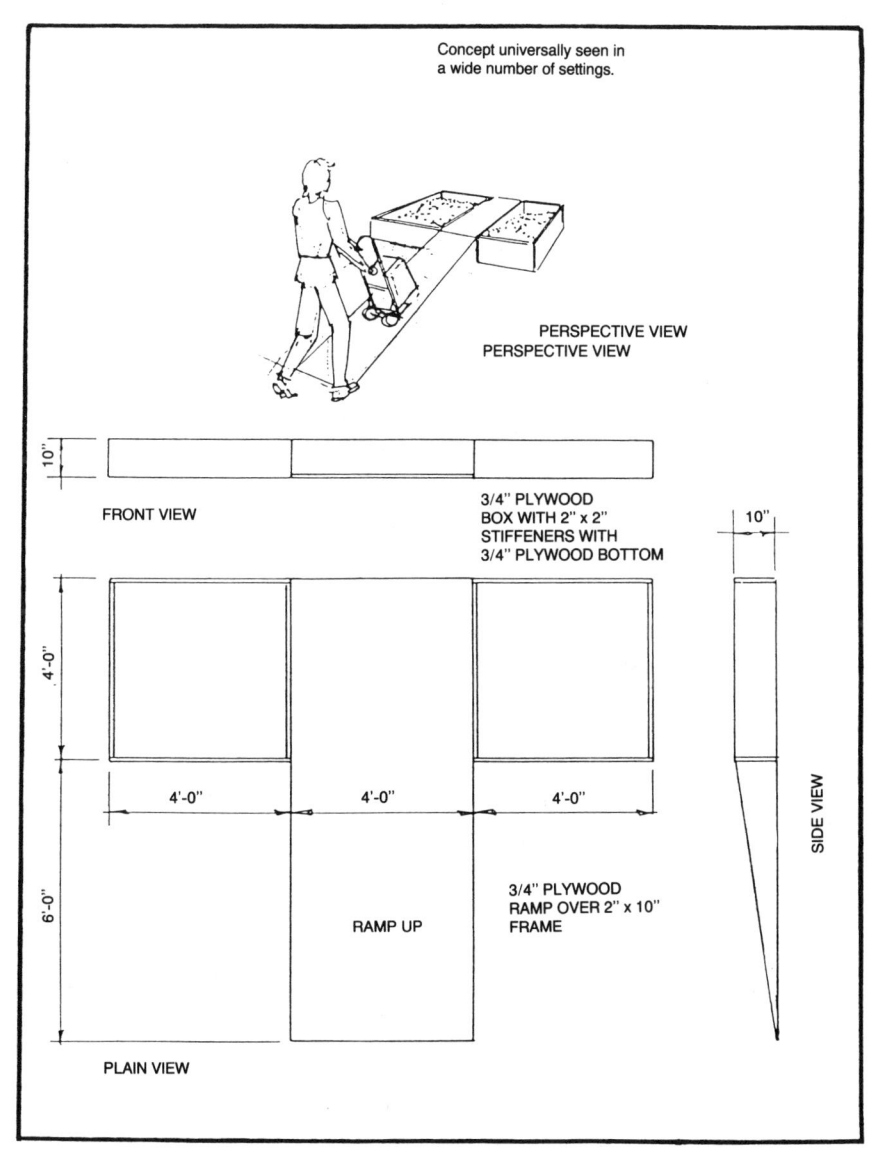

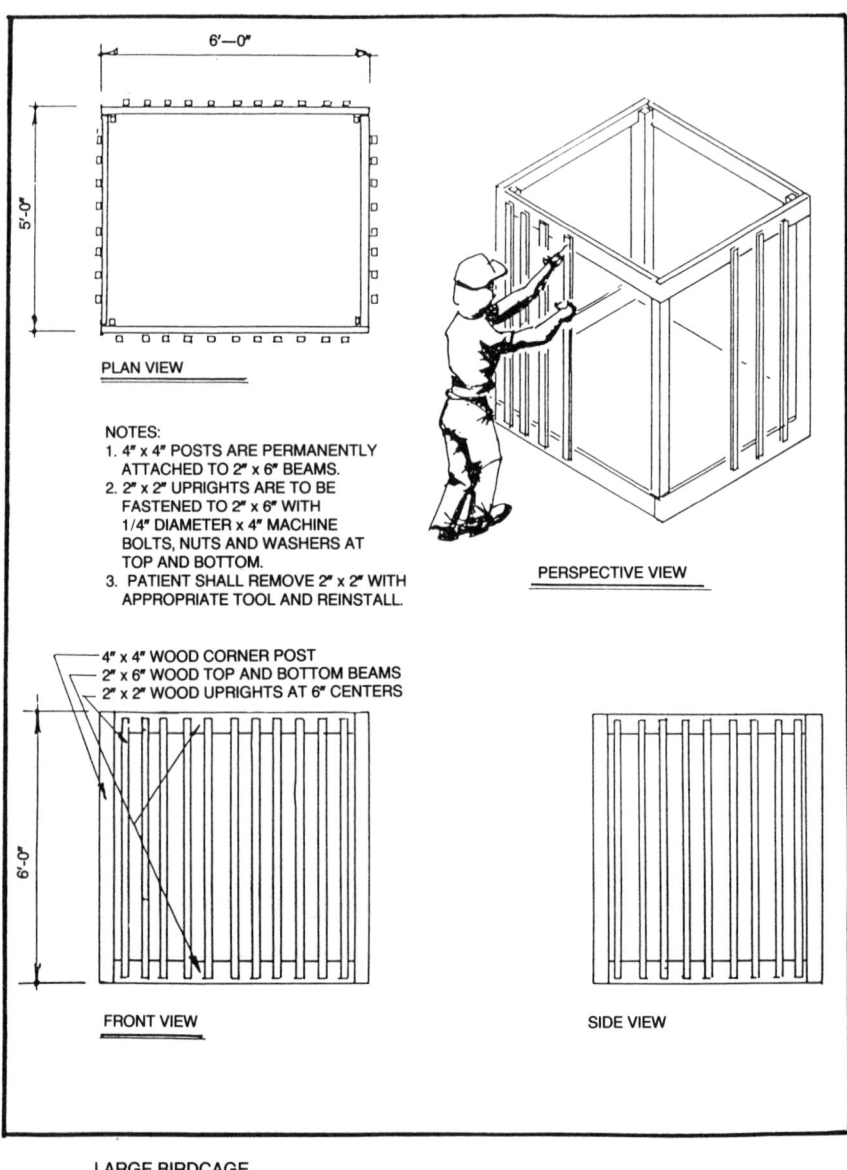

222  WORK INJURY

**Figure 15-10** Large birdcage work simulation. *Source:* Reprinted from *Work Capacities' Technical Training Manual* (p. 382) with permission of Work Capacities Inc. © 1986.

discharge from the program. To recommend discharge with release to return to a specific job, the rehabilitation team must be able to document the client's ability in relation to the job requirements. For those clients with no specific job to return to, the team must demonstrate either achievement of work-related rehabilitation goals or a plateau in progress.

*Comparative Functional Analysis*

A comparative functional analysis should be performed before a client is discharged from the program. This is simply a reevaluation of those areas addressed on the base line functional analysis. It provides clear documentation of progress and enables establishment of recommendations regarding physical demand characteristics of future jobs.

On discharge clients must clearly understand their current capabilities and limitations and understand that although they are not yet perfect, their physical and emotional condition is "good enough" for successful reentry into the work force.

## PROFESSIONAL APPLICATION

Certainly, each type of work hardening model described has its place in rehabilitation of the worker. However, certain generalizations can be made regarding the development of new programs or the evaluation of existing programs:

1. Excellent communication involving the center or program, the injured worker, referral sources, and the staff is vital.
2. Specific admission and discharge criteria must be adhered to, especially when referrals involve general therapy departments and their primary care patients. Criteria also help physicians, carriers, and other referral sources understand what is different about work hardening compared with other resources and help define the caseload to be appropriate to your center's philosophy and methods.
3. The concept of "good enough" must be understood by staff. The patient-worker will probably advance sufficiently in work hardening to return to work before he or she is capable of doing all activities of daily living. As this person continues to use his or her improved state, more skills of a general type may return and extend beyond work parameters.
4. The goal is to get the patient-worker into a self-responsibility versus disability mindset as quickly as possible; this is partially accomplished

through a trust relationship. Not only team members but also a case manager can frequently facilitate this end.
5. Finding an effective means of comparing job to functional performance level is a very important factor in work hardening and one requiring constant exploration for better methods.
6. Obtaining a base line analysis before beginning a work hardening program, and comparing it later with an equivalent reevaluation format, offers an excellent data base for statistics to support a program's effectiveness.

## ORGANIZATIONAL PROBLEMS

Clearly, a multidisciplinary team approach is well suited to manage the multitude of problems facing the injured worker. However, teams are only effective if they are well coordinated and managed. Inherent structural or departmental characteristics of organizations do not always facilitate this coordination and management. Care must be taken to avoid certain problems that arise when a work hardening team, made up of members from various departments, is assembled:

1. Communication among and between team members may be a problem. Irregular or ineffective communication impacts on reports, which appear fragmented. Frequently, each member may have his or her own concept of what work hardening should accomplish.
2. Case management is difficult if all team members are not synchronized. Such problems impact on the worker-patient and any vested party who needs to understand case process and cannot find a single source to convey it.
3. There may be inadequate time for planning, implementation, and scheduling, if various team members have conflicting time responsibilities.
4. If work hardening is not in one location, there may be difficulty in service integration.

## ENVIRONMENTAL PROBLEMS

Although the work hardening models address many return-to-work problems, there continue to be problems that exist, and as yet are unsolved, because of environmental conditions.

The best case outcome from a cost and success standpoint is to return the worker to his or her former job. However, there is very little data on whether returning the worker to the same job may result in a greater chance of re-injury. Statistics indicate that a worker who returns to work is more likely to re-injure himself or herself or aggravate a former injury than he or she was to injure himself or herself before the incident. The unanswered question remains, Are we truly doing the worker a service when we encourage him or her to return to his or her former job?

When a client is unable to return to the former job, other issues become paramount. The biggest problem revolves around the financial implications of the new job. The "typical case" involves a client who has performed heavy labor his entire life and has not been trained or educated for other work. With relatively few skills he finds himself faced with a dramatic cut in pay when he obtains a new job. Coupled with the fact that he may be receiving other benefits while disabled (such as credit disability policies that make his house and car payments) there is very little incentive to return to gainful employment.

Another factor to consider is employer prejudice; that is, once a worker has been "disabled" it is difficult to find employment with that disability. Also, if a person has restrictions, an employer may pass over that person rather than deal with the restrictions.

The final problem lies with the system. All of the concerned parties have separate goals in each compensation case. (See Table 15-1.) The worker wants to regain his or her pre-injury status, even though this may not be realistic. The insurance company wants to settle the case as inexpensively as possible. The employer wants to know where he or she stands and the physician wants to please the patient. When an attorney is involved, the case becomes even more complex because the goal is to obtain the largest settlement possible. Some of these goals are in direct opposition to one another and the client is in the middle. This tends to create a frustrating, complicated situation that is very difficult to resolve.

## CASE CLOSURE

Many situations are difficult to resolve and work hardening centers must realize they cannot "fix" everything. The major role that a center can fill, however, is to expose the issues, deal with those that can be resolved and realize when the outcome is good enough. If possible, all parties involved need to reach the point where all of the issues are on the table and the best for all concerned is to go ahead and resolve the case and get on with life. The effective work hardening center many times can expedite that process thereby moving the case toward resolution.

**Table 15-1** Return-to-Work Delay Issues and Work Capacity Answers

| Vested Party | Problems/Needs | Work Capacity Services |
|---|---|---|
| The Worker | 1. Prolonged recovery<br>2. Disability mindset<br>3. Secondary problems<br>4. Distrust of the systems:<br>• Medical<br>• Employer<br>• Carrier<br>5. Family alienation | 1. Decrease primary care therapy time; earlier referral begins appropriate rehabilitation sooner.<br>2. Engage client in purposeful activity to change belief system and physical limitations.<br>3. Promote sense of trust through interaction with work hardening staff. Staff act as liaison (case manager) between worker and the physician, employer, and carrier.<br>4. Educate family members through interaction and enlist them as at-home positive reinforcers. |
| The Physician | 1. Safeguard patient<br>2. Return-to-Work release:<br>• Arbitrary opinion and caution by physician<br>• Subjective report from patient<br>3. Search for appropriate therapy/rehabilitation<br>4. Limited options with symptomatic return | 1. Provide the physician with objective test data that support his or her other orders.<br>2. Tell the physician the injured worker has demonstrated safe, productive work behavior and biomechanical performance.<br>3. Inform the physician about the actual workplace and the job in particular.<br>4. Provide the physician with all information within a professional and highly specialized report format.<br>5. Offer the physician an excellent means of avoiding physical and mental/emotional secondary problems in the patient-worker or remediating them effectively through appropriate rehabilitation services.<br>6. Provide ongoing communication between case manager and physician in an efficient manner. |

**Table 15-1** Continued

| Vested Party | Problems/Needs | Work Capacity Services |
|---|---|---|
| The Employer | 1. Light duty complications:<br>• No temporary light duty; prolonged return<br>• Inappropriate light duty. Will it reinjure the worker?<br>2. What modifications to make<br>3. Want to help but concerned about what substitute job can be offered<br>4. Self-insured companies:<br>• Where to begin with rehabilitation services<br>• Cost of rehabilitation services | 1. Upgrade worker sufficiently through work hardening to eliminate need for light duty or temporary assignments. Worker returns to work earlier.<br>2. Change status to fully functioning employee through returning to regular work position and normal productivity rate.<br>3. Provide information regarding tool or work station modification from a work hardening standpoint.<br>4. Can work with self-insured companies for a system that provides referrals and measures for cost efficiency. |
| The Carrier | 1. Commitment to control costs of a workers' compensation case<br>2. Effective rehabilitation resources; good communication | 1. Provides prediction for carrier regarding costs, time length, and outcome of rehabilitation.<br>2. Provides excellent communication through case manager contact. |

## SUMMARY

Work Hardening is a concept whose time has come. Its major function is to assist with case resolution and maximal rehabilitation in the most expedient and cost-effective manner possible.

There are many types of programs emanating from various kinds of settings. The future challenge for professionals in the work hardening arena is to improve the ability to predict who can be helped and when and then to deliver what was promised. In doing so we can offer the highest quality services in the most cost-efficient manner.

### REFERENCES

1. Matheson LN, Ogden LD, Violette K, et al: Work hardening: Occupational therapy in industrial rehabilitation. *Am J Occup Ther* 1985; 39:314-321.
2. Harvey-Kretting L: The concept of work in occupational therapy: A historical review. *Am J Occup Ther* 1985; 39:301-307.
3. Marshall EM: Looking back. *Am J Occup Ther* 1985; 39:297-300.
4. Kemp B, Kleinplatz F: Vocational rehabilitation of the older worker. *Am J Occup Ther* 1985; 39:322-326.
5. Sheehan H, Viesko B: Work therapy in an army hospital. *Am J Occup Ther* 1958; 12:176, 188-189.
6. Wilke NA, Sheldahl LM: Use of simulated work testing in cardiac rehabilitation: A case report. *Am J Occup Ther* 1985; 39:327-330.
7. Fitzler SL: Attitudinal change: The Chelsea back program. *Occup Health Saf* 1982 (February):24-26.
8. Llorens LL, Levy R, Rubin EZ: Work adjustment program: A prevocational experience. *Am J Occup Ther* 1964; 28:15-19.
9. Symington DC, Lewis MP: Cost benefit analysis of a hospital and university based prevocational program. *Arch Phys Med Rehabil* 1978; 54:341-347.
10. Reuss EE, Rawe DE, Sundquist AE: Development of a physical capacities evaluation. *Am J Occup Ther* 1958; 12:1-8, 14.
11. Creighton C: Three frames of reference in work-related occupational therapy programs. *Am J Occup Ther* 1985; 39:331-334.
12. Holmes D: The role of the occupational therapist—work evaluation. *Am J Occup Ther* 1985; 39:308-313.
13. Howe MC, Weaver CT, Dulay J: The development of a work-oriented day center program. *Am J Occup Ther* 1981; 35:711-718.
14. Rosenberg B, Wellerson T: A structured prevocational program. *Am J Occup Ther* 1960; 14:57-60, 106.
15. Nachemson A: Work for all—for those with low back pain as well. *Clin Orthop* 1983; 179:77-85.
16. Bettencourt CM, Carlstrom P, Brown SH, et al: Using work simulating to treat adults with back injuries. *Am J Occup Ther* 1986; 40:12-18.

17. Mattingly S: Rehabilitation of registered dock workers at Garston Manor. *Proc R Soc Med* 1971; 64:757-760.

18. Newman CE Jr: Rehabilitation can be a cost containment device. *Hospitals* 1979 (March):45-46.

19. Mayer TG: Rehabilitation of the patient with spinal pain. *Orthop Clin North Am* 1983; 14:623-637.

20. Reggio WA: Some thoughts on occupational therapy. The workshop and rehabilitation. *Am J Occup Ther* 1948; 2:164-165.

21. Schaeffer JN: Factors in a total rehabilitation program for low back pain. *J Occup Med* 1967; 9:12-15.

# Part VII

# Functional Restoration of the Patient with Chronic Spinal Disorders

Spinal injuries, particularly low back injuries, have historically perplexed the medical profession. In the absence of clearly diagnostic objective testing, patients have continued to complain of debilitating pain and to suffer chronic dysfunction. Many times continued and varied medical intervention has produced little relief of symptoms or increase in function.

Likewise, the employer has been frustrated in attempts to return the injured worker with chronic low back pain to work. Return to work is often associated with increased discomfort and subsequent lost time from work. In some cases, re-injury occurs with no better recovery results from the re-injury than those from the first injury.

The patient with spinal dysfunction also feels the frustration of physical pain and loss of work ability. There are additional negative feelings toward the medical practitioner who cannot effect a cure, and toward the employer whose job is the potential site of more pain.

### Section 16  Functional Restoration of the Patient with Chronic Spinal Disorders

A comprehensive, multidisciplinary approach to rehabilitation of the worker with spinal injury is presented. Without seeking a cure or using invasive medical techniques, concentrated functional restoration is centered on work-related gains. The program success has made a strong impact on the way in which treatment of the patient with spinal dysfunction is directed.

# Section 16

# Functional Restoration of the Patient with Chronic Spinal Disorders

*Tom G. Mayer*

## IS MEASUREMENT THE KEY TO REHABILITATION OF SPINAL DISORDERS?

Consciously or unconsciously, measurement is a part of all activities of daily living. The senses are used regularly to determine temperature, distance, size, or texture. In many cases, a fairly qualitative evaluation will suffice, usually because there is no visual or tactile limitation on frequent repeated assessment of the object of interest, such as observing a moving vehicle in traffic or placing food in the mouth. In some cases, however, greater precision is necessary, usually occurring when the object at hand is *not* amenable to regular visual inspection. In these cases, a quantitative assessment usually is accompanied by a knowledge of a range of normal values (obtained by testing many "normal" subjects). Variation from a mean score in a patient then is used as a mechanism for evaluating the presence of disease or dysfunction or the return to an improved state from an abnormal one. The spinal anatomy does not lend itself easily to visual or tactile examination and thus demands indirect, quantitative measurement to describe its performance.

Musculoskeletal clinicians are somewhat spoiled by visual access to most parts of the organ system. They have not had to develop the discipline of the hematologist and the cardiologist who must rely on indirect measurements (eg, blood components, heart rate, and blood pressure) in the diagnosis of abnormalities. Although quantitative measures of joint motion or muscle strength have been available for many years, all too often a hasty qualitative assessment is substituted for the quantitative one. The novice physician or therapist soon learns that quantitative measurements are time consuming and generally unnecessary for routine clinical practice. The comparison of motion between both knees and the size of a quadriceps (often without even using a tape measure) is "eyeballed," reserving the use of a goniometer or isokinetic testing device for special occasions when an evaluation specifies quantification.

In the spinal anatomy, the small, inaccessible, three-joint complexes stacked on each other cannot be inspected easily. Intersegmental movement in the spine is difficult to measure even with biplanar x-ray devices. Multiple small muscles interdigitate over variable numbers of segments, and ligamentous structures may share surprising amounts of load in certain joint positions. Furthermore, bilateral comparisons are impossible. Until recently, there were no valid indirect measurement methods available to assess function of the spine, which led to ignorance of pathological processes in the vast majority of cases of spinal dysfunction not resolving spontaneously. Although methods for direct visualization still are not available, novel technology for assessing spinal function has become part of clinical routine. Yet, many physicians persist in ignoring or refusing to use such technology. In so doing, therapeutic errors are encouraged, outcomes remain unevaluated, and fringe or fad treatments are perpetuated.

The weaker clinical areas must be recognized before the discipline of rehabilitation can move forward. In diagnosis of spinal dysfunction, lack of recognition of the deconditioning syndrome has adversely affected therapeutics. Many people currently embracing work hardening (often as fervently as they previously embraced pain management) continue to do so with eclectic therapies applied uniformly to all patients, rather than individualized on the basis of functional testing. Surgery is performed on only 2% to 3% of patients with spinal disorders. Nevertheless, surgeons will search diligently for that small percentage with a wide variety of expensive and sophisticated diagnostic tools (eg, computed tomography scan, magnetic resonance imaging, electromyography, myelography). A structural diagnosis, when made, may lead to the ability to "fix" an anatomic aberration, such as a prolapsed disk. For the remaining 97% of the back-injured population, spontaneous recovery may account for many "cures." However, a substantial percentage of patients, perhaps as high as 30% to 40%, will show some evidence of disuse and deconditioning, making them candidates for physical retraining once the functional deficits are identified. Without quantification, the deficits are simply not recognizable, leading to inevitable overutilization or underutilization of therapeutic services as a result. This observation is not merely true for spinal disorders, nor has it escaped the attention of health care planners. Medicare requires periodic testing to document progress in other areas of rehabilitation. It is likely that similar rules will ultimately apply to treatment of spinal problems, once their necessity becomes more generally perceived.

## HOW CAN EFFICIENT SPINAL MEASUREMENTS BE DESIGNED?

The term *deconditioning syndrome* has been applied to the cumulative disuse changes produced in the chronically disabled patient suffering from spinal dysfunction. It is initially produced by the immobilization and inactivity attendant on injury. It is supplemented by disruption of spinal soft tissue and scarring resulting from a surgical approach or repetitive microtrauma. As pain perception is enhanced, learned protective mechanisms lead to a vicious cycle of inactivity and disuse. As physical capacity decreases, the likelihood of fresh sprains and strains to unprotected joints, muscles, ligaments, and disks increases. These inevitable alterations of pain and function are perceived by the patient as a recurrence or reinjury.

In spinal function assessment, experience has been drawn from assessment of the extremities in identifying elements of performance that are of value in characterizing extremity physiologic "functional units." Such factors as range of motion, strength, neurological status, endurance combined with whole body aerobic capacity, and activities of daily living measurements, are traditionally assessed.

Unfortunately, evaluation of extremity neurological function (eg, straight leg raising, lower extremity strength, sensation and reflexes in dermatomal/myotomal patterns) is still viewed by the majority of clinicians as the ideal spine functional evaluation. This is simply another test to find the 2% to 3% of patients requiring surgical intervention.

As tests of spinal function, these neurological characteristics may be irrelevant for several reasons. First, they are a measure of acute change when noted in relation to surgical pathology. In the chronic situation, persistence of neurological changes generally reflects epidural fibrosis or other permanent, noncorrectable anatomic abnormalities. In addition, the neurological deficits, though emanating from spinal structures, are perceived by the patient as extremity abnormalities producing pain, sensory changes, and weakness of arms and legs. Once appropriate surgical decompression has been achieved, rehabilitative treatment of a "drop foot" is focused on bracing, mobilizing, and strengthening the foot and ankle, not on retraining the spinal elements. In sum, what the clinician currently views as standard objective functional tests may provide no useful information to overcome spinal deconditioning.

Isometric lifting tests have been used to measure back strength.[1,2] Although engineering principles require that the force of an isometric test be transmitted through the body from hands to foot/floor contact, the spine often can be placed in such a position that virtually no spinal muscular strength is required to support the load.[3] This exemplifies the first criterion for any functional evaluation: the test must be relevant to the physiology being measured. If isolated spinal muscular strength is to be assessed, then measures of trunk torques in flexion/extension, rotation, or lateral bend must be sought, not whole body tasks such as lifting.

A second critical principle of measurement is the need to know the validity and reproducibility of the measurement. Validity refers to the accuracy of the test device, whereas reproducibility refers to the ability of the test (device *plus* subject) to give a repeatable and precise measure of a clinical variable. A valid test is not necessarily reproducible and vice versa. An invalid test is simply useless, but an irreproducible test may reflect actual clinical reality (such as comparing intersubject body weight or changes in spinal mobility before and after exercise in the same person). As in these examples, reproducibility problems often can be corrected by restructuring the test protocol.

Once it has been determined that a functional test is valid, reproducible, and relevant, an effort factor must be defined. Without the ability to identify suboptimal effort, invalid low readings may be accepted as true physical deficits. Although suboptimal effort may reflect a clinical abnormality (eg, low motivation, pain, fear of injury, or a personality disorder leading to conscious malingering), the clinician must be able to assess whether he or she is dealing with a true physical deficit or not.

Finally, once the appropriate device and test protocol have been chosen, the hard work begins. A normative data base must be compiled on a large sample to allow for comparisons to a patient population. Degree of deviation from "normal" may significantly affect treatment protocols. Although much more should be said about measurement criteria, it is most important to select the optimal measurement device first to avoid the onerous task of repetitively collecting normative data.[1,4-10]

Quantified functional evaluations available for the lumbar spine include inclinometric range of motion[7,8,10]; trunk strength[4-6,11], aerobic capacity measurement[9], lifting capacity[1,2,12,13]; and standardized task performance capacity.[14] A detailed discussion of these measurements is beyond the scope of this chapter. The most expensive devices are those used to assess isokinetic strength and lifting capacity. Isokinetics has the advantage of providing a measurement of dynamic performance with a method that "locks in" the speed and acceleration variables so that they become known quantities. Then, torque or force become the only independent variables, making calculation of both interindividual and intraindividual differences relatively easy. Other dependent variables, such as work and power, can be derived from curves produced on an isokinetic device. On the other hand, when these variables are not stabilized, reproducing a test can become merely a computer fabrication. While the Productive Rehabilitation Institute of Dallas for Ergonomics (PRIDE) had the privilege of working with the original Cybex prototypes (Lumex; Ronkonkoma, NY) to develop many of the trunk strength testing concepts, other manufacturers also produce isokinetic trunk strength testing equipment. They include: Lido (Loredan, Inc; Davis, Calif), Kin-Com (Chatteck Corp; Chattanooga, Tenn), and Biodex (Biodex Co; Shirley, NY). Other functional measurements (eg, range of motion), however, can be performed with equipment that is considerably less expensive.

## WHAT BARRIERS TO FUNCTIONAL RECOVERY MAY ARISE?

A multitude of psychosocial and socioeconomic problems may confront the patient recovering from a spinal disorder, particularly if disability from a productive life style is associated with the back pain. The patient's inability to see a "light at the end of the tunnel" may produce a severe situational depression often associated with anxiety and agitation. The back injury may be a sign of emotional conflicts involving rebellion against authority or job dissatisfaction.[15-17] Personality changes may be manifest in noncompliance or in anger and hostility directed toward the therapeutic team. Minor head injuries; organic brain dysfunction from age or the effects of alcohol or drugs; or limited intelligence may produce cognitive disorders that make patients difficult to manage and refractory to education. A variety of personality disorders such as sociopathy also may complicate treatment.[9,10,18-20]

Many chronic spinal disorders exist within a "disability system." Workers' compensation laws were initially devised to protect workers' income and provide timely medical benefits after industrial accidents. Employers ultimately agreed to this because of a compensatory benefit: in return for providing these workers' rights, employers were absolved of certain consequences of negligence, generally including cost-capped liability for any injury, no matter how severe, and set by state statute. As in any compromise situation, certain disincentives to rational behavior may emerge. One outcome of a guaranteed paycheck while temporary total disability persists is that there may be no clear incentive for an early return to work. A casual approach to surgical decision making and rehabilitation may lead to further deconditioning, both mental and physical, thus making ultimate recovery more problematic. Complicating matters even further is the observation that no group (other than the employer) has a verifiable financial incentive to return patients to productivity as soon as possible. In consequence, an odd assortment of health professionals, attorneys, insurance company personnel, and vocational rehabilitation specialists may be minimally motivated to combat foot-dragging on the disability issue. Altering the contingencies may correct some of the problems. However, this assumes that the present system has not already evolved to a near-perfect balance of interests, or that the legislators will respond to changes in outlook regarding optimal patient care.

## FUNCTIONAL RESTORATION: REBUILDING THE DECONDITIONED MIND AND BODY

Generally, nonsurgical treatment for musculoskeletal injury is divided into two phases: conservative care and rehabilitation. Conservative care is commonly passive, relying frequently on modalities and rest, which are intended to promote healing. Rehabilitation is usually believed to be more active, relying primarily on education and exercise.

The passive modalities used in conservative care for the spine injury fall into six categories:

1. Rest (bed rest, immobilization, braces).
2. Thermal agents (heat, cold, ultrasound, diathermy).
3. Mechanical agents (massage, traction, manipulation).
4. Acupuncturoid agents (acupuncture, transcutaneous electronic nerve stimulator [TENS], T-CET).
5. Pharmacologic agents
   - Invasive (epidural, facet, trigger point, sacroiliac injections).
   - Noninvasive (medications such as opiates, tranquilizers, anti-inflammatories, antidepressants).
6. Miscellaneous or fringe agents ("activator" guns, craniosacral therapy).

Although these treatments are often very much appreciated by the patient as a sign of the clinician's caring and nurturing skills, their effect has been antithetical to the intended purpose if used too frequently or for too long. Immobilization obviously encourages disuse, but passive treatment of any kind can likewise produce reliance on health professionals and dependency inconsistent with the goal of reactivation of a productive citizen at the conclusion of treatment.

Pain remains a subjective phenomenon, the extent of which is really known only to the person experiencing it. Many discussions of secondary gain involved in pain self-report have produced a widespread skepticism among health professionals regarding the patient with chronic pain. This author believes that the health professional has a duty to the patient to believe what the patient says, but to place that description in appropriate context of the patient's overall behavior. When all has been done to passively reduce the patient's pain and the pain can be reduced even further by a program of progressive exercise and education, the patient should be expected to participate. Avoidance of participation or attempts to sabotage treatment associated with magnified and incessant complaints of pain should lead the experienced clinician to suspect psychosocioeconomic barriers to functional recovery. If the health professional, in conjunction with other members of the disability system, can find a way to address these issues, the patient may yet respond to treatment. Statistically, even among lost time injuries in industry (both personal and work related), more than 90% of patients are back at work within 3 months. Of the group remaining, however, a majority will still be disabled at the end of 1 year. Thus, the greatest attention should be directed toward the group who have been disabled more than 3 to 4 months. Although disuse changes may occur as early as 6 to 8 weeks after an injury, it is the group subjected to disuse for more than 4 to 6 months, or who have had delayed or multiple surgeries, who will have a much higher incidence of functional loss.

The approach to the patient with quantifiable functional deficits is predicated on involvement of barriers to functional recovery. If these factors are minimal contributors to disability, or if the patient has decided to continue working despite pain, treatment usually can be restricted to physical functional restoration including reconditioning, job simulation, and/or work hardening without the need for extensive psychological or educational components. However, if nonorganic factors are present to a great extent, a period of intensive treatment, combining physical and psychological rehabilitation, usually will be necessary to obtain an effective outcome. Such treatment must truly be through a team approach, because no one specialty can provide all of the knowledge required to achieve adequate outcomes in a comprehensive treatment program. At PRIDE, the team comprises physicians, nurse managers, physical therapists, occupational therapists, psychologists, and vocational rehabilitation specialists. Many of the roles of these therapists are nontraditional and are based on acquiring new skills in understanding quantification of function technology and psychosocioeconomic barriers to functional recovery. Most important, a team approach with excellent internal and external communication is necessary to achieve optimal results.

The primary eligibility factor for beginning a functional restoration program is the prior determination of need for an additional surgical procedure. Surgery should be medically "ruled out" (as much as possible) before commencement of functional restoration. This does not mean that surgery is never performed after attempts at functional restoration. In fact, if high levels of physical capacity are achieved and the patient remains significantly symptomatic, additional structural testing to identify previously overlooked surgically treatable pathology may be indicated. However, in most cases, the effort and expense of intensive functional restoration training cannot be justified in a patient not adequately investigated or simply biding time until surgery. In the circumstance in which a patient with psychosocial problems is being somewhat reluctantly considered for surgical intervention, a short period of outpatient training and counseling may be effective in clarifying motivational factors and giving some insight into the patient's probable postoperative course.

Once surgery has been ruled out as an alternative, the stage is appropriately set for functional restoration. At that point, all indicated diagnostic testing, invasive therapies (therapeutic injections, medications) and/or passive therapy modalities should have been offered. Usually several months, or even years, will have elapsed. If the patient remains symptomatic, and if a quantified evaluation of functional capacity demonstrates sufficient physical deficits, a rehabilitation program is indicated. The actual mechanics of such a program are complex and involve several phases. There is not only "one right way" to perform the training. At PRIDE, the physical therapy department focuses its activities on rebuilding the specific injured area of the body; that is, the lumbar (or cervicothoracic) spine "functional unit." After initial quantitated evaluation, the degree of deficit determines the starting point for the various mobility, strengthening, and aerobic exercises, as well as which exercise which will receive greater emphasis. In general, mobility is restored first before strength and aerobic capacity/endurance issues are addressed. An exercise tolerance test may be indicated before embarking on intensive aerobic training in certain risk groups.

In contrast to the responsibilities of the physical therapists, the PRIDE facility has occupational therapists supervising restoration of functional task performance capability; that is, total body tasks such as lifting, bending, and carrying. This involves the coordination of several body "functional units" to produce greater positional and activity tolerance. Return to regular work activities through work simulation, education, and counseling may assist in the process. Training in this area involves learning to perform specific activities such as twisting, climbing, bending, and carrying. These are done initially in isolation and then in combination with other activities to build endurance, coordination and agility. Work hardening is simply a term for programs promoting the increased tolerance to prolonged activity resulting from a supervised training program.

Psychologists participate in the team approach with a multimodal disability management program that features a cognitive-behavioral treatment orientation to help patients separate pain from disability. The psychological treatment program depends on an extension of the quantification of function approach, using predictor personality tests, to the psychological realm. The program is "crisis-oriented" to deal with maladaptive behavior and psychosocial issues (both present and future) with which the patients have to cope. It does not engage in prolonged psychotherapy to alter premorbid maladaptive mechanisms with which the patient has been troubled, but with which he or she has coped successfully in the past. Dealing with these issues delves into psychopathology that did not interfere with the patient's work before injury, and also dilutes scarce treatment resources.

Combined with training is an educational program directed at specific outcome criteria. The educational program may be provided by therapists, nursing personnel, or rehabilitation counselors, but in this author's facility it is provided primarily by psychologists. Classes focus on the various issues of pain and disability that have already been discussed. Even the outpatient, less intensive program is much more comprehensive than the usual "back school" training in this regard. Psychologists may be responsible for such traditional areas as stress management, but also provide their ability to understand and communicate on the functional barriers in a cognitive-behavioral educational program, including coping skills, assertiveness training, and training on rational versus emotional responses. For the disabled patient with spinal disorders, the treatment process represents a culmination of a crisis in their lives and all appropriate educational opportunities should be provided to them. On the other hand, dependency-producing treatment, such as prolonged psychotherapy, should be avoided.[9]

For the patient with subacute, postoperative or chronic episodic pain who is working without substantial psychological impairment or financial disincentives, a simple outpatient program may provide sufficient rehabilitation. In such a program, patients may undergo repeated quantitative evaluations of physical function capacity, after which a determination is made as to the need for continued treatment. Patients are continued in such an outpatient program until normal physical capacity is achieved, a plateau is reached, or symptoms abate (if this is the patient's only goal). The physician must help to make decisions on treatment and address questions of patient motivation. Ultimately, however, the quality of the therapists' training and their ability to apply principles of functional quantification determine program success.

Once all passive medical treatment has been exhausted, and the patient has been given maximum opportunity for active physical functional restoration, the medical care system has done all that it can. The patient is then "medically stationary" and determination of functional limitations and impairment can be made to identify the patient's work capabilities and to resolve outstanding legal and financial compensation issues. The well-motivated person can be restored to high levels of functional capacity, just as many football players who have undergone multiple surgeries for a knee injury have returned to professional sports after ideal "sports medicine" rehabilitation. It should be clear that the conclusion of a functional restoration program allows a return to some type of work for almost all people, as well as more rapid resolution of litigation. Although the medical and legal systems will continue to give appropriate credence to patient self-reports of pain and disability, it is likely that in the future greater reliance will be placed on the diagnosis of the structural injury sustained and the degree of effort/physical capacity produced by the claimant in achieving the highest possible levels of functional capacity despite pain and scarring.

The PRIDE intensive program lasts 3 weeks and may occasionally be shortened to 2 weeks ("light duty" working patients). Follow-up varies up to 6 weeks and depends on the job demands and job availability, as well as the degree of functional capacity deficits and psychosocial deconditioning exhibited by the patient. Follow-up decisions usually are made only in the intensive phase of the program. However, the known range of treatment options allows a third party payer to evaluate the maximum and minimum ranges of treatment costs.

A goal-oriented approach is used during the 3-week comprehensive program at PRIDE. The patient departs with a specific follow-up plan, including duration of additional treatment and work goals. Job demand categories from 1 (sedentary to light) to 4 (very heavy) have been modified from existing sources and assist therapists in occupational counseling.[9] External communication with employers, attorneys, insurance companies, and vocational rehabilitation agencies assures optimum follow-through in resolving issues of work, compensation, job modification, and retraining.

Outcomes of such a program have been measured and are well documented in the literature.[10,21,22] A return to work rate of more than 80% of patients completing a comprehensive program can be anticipated. Such results can be expected to be sustained through a 2-year follow-up, with re-injury rates for the treated group no higher than that of the general working population. The need for additional medical, surgical, physical therapy, and chiropractic treatment was two to five times greater in a comparison group of patients denied functional restoration treatment than in the group completing treatment. Certain groups, such as those who "dropped out" during the intensive treatment program or who were receiving Social Security disability income had particularly dismal outcomes and require very careful attention. Outcomes have not been measured for the noncompliant group (representing about 15% to 20% of referrals) of chronic spinal disorder patients who simply refuse to indulge in any treatment program requiring their active participation.

## HOW CAN THESE CONCEPTS BE SUMMARIZED?

The focus of this chapter has been to convey the importance of measurement of physical capacity in devising a treatment program. Quantification of function through indirect means is the only means available to document progress in a treatment program, to inform the patient and clinician of physical deficits and the effect of treatment, identify effort, and to provide objective work evaluation information. Ultimately, after appropriately validated prospective studies have documented their predictive value, such functional tests (or subsets) may be used for worker selection in such areas as pre-employment screening or placement activities. If used fairly and scientifically, such techniques may offer the opportunity of better matching the work force to the task, combined with modifying the task to suit the average worker. Significant implications for disability and impairment evaluation also are inherent in functional quantitative measurements. Using these measurements, rehabilitation programs, known as functional restoration, can be derived employing a "team approach" to provide optimum therapeutics. Costs rise as more intensive treatment is provided; but if appropriate selection and effective early conservative care is applied, the cost/benefit ratio for such therapy will remain low. In this way, the sports medicine approach to functional restoration addresses the critical personal and societal issues that must be faced in dealing with the chronically disabled patient with structurally irremediable spinal disorders.

**REFERENCES**

1. Kishino N, Mayer T, Gatchel R, et al: Quantification of lumbar function part 4: Isometric and isokinetic lifting simulation in normal subjects and low back dysfunction patients. *Spine* 1985; 10:921–927.

2. Chaffin D, Herrin G, Keyserling W: Pre-employment strength testing: An updated position. *J Occup Med* 1978; 20:403–408.

3. Gracovetsky S, Farfan H: The optimum spine. *Spine* 1986; 11:543–573.

4. Smith S, Mayer T, Gatchel R, et al: Quantification of lumbar function part 1: Isometric and multi-speed isokinetic trunk strength measures in sagittal and axial planes in normal subject patients. *Spine* 1985; 10:757–764.

5. Mayer T, Smith S, Keeley J, et al: Quantification of lumbar function part 2: Sagittal plane trunk strength in chronic low back pain patients. *Spine* 1985; 10:765–772.

6. Mayer T, Smith S, Kondraske G, et al: Quantification of lumbar function part 3: Preliminary data on isokinetic torso rotation testing with myoelectric spectral analysis in normal and low back pain subjects. *Spine* 1985; 10:912–920.

7. Keeley J, Mayer T, Cox R, et al: Quantification of lumbar function part 5: Reliability of range of motion measures in the sagittal plane and an *in vivo* torso rotation measurement technique. *Spine* 1986; 11:31–35.

8. Mayer T, Tencer A, Kristoferson S, et al: Use of noninvasive techniques for quantification of spinal range-of-motion in normal subjects and chronic low-back dysfunction patients. *Spine* 1984; 9:588-595.

9. Mayer T, Gatchel R: *Functional Restoration for Spinal Disorders: The Sports Medicine Approach to Low Back Pain.* Philadelphia, Lea & Febiger, to be published.

10. Mayer T, Gatchel R, Mayer H, et al: A prospective two-year study of functional restoration in industrial low back injury: An objective assessment procedure. *JAMA* 1987; 258:1763-1767.

11. Mayer T: Physical assessment of the postoperative patient in failed back surgery, in White A (ed): *State-of-the-Art Reviews.* Philadelphia, Hanley & Belfus, 1986, pp 93-101.

12. Troup J, Foreman T, Baxter C, et al: The perception of back pain and the role of psychophysical tests of lifting capacity: 1987 Volvo Award in Clinical Sciences. *Spine* 1987; 12:645-657.

13. Mayer T, Barnes D, Kishino D, et al: Progressive isoinertial lifting evaluation, part I: A standardized protocol and normative database. *Spine* 1988 (in press); Mayer T, Barnes D, Nichols G, et al.: Progressive isoinertial lifting evaluation, part II: A comparison with isokinetic lifting in a disabled chronic low back industrial population. *Spine* 1988 (in press).

14. Mayer T: Using physical measurements to assess low back pain. *J Musculoskel Med* 1985; 2:44-59.

15. Bigos S, Spengler D, Martin N, et al: Back injuries in industry: A retrospective study II. Injury factors. *Spine* 1986; 11:246-251.

16. Bigos S, Spengler D, Martin N, et al: Back injuries in industry: A retrospective study III. Employee-related factors. *Spine* 1986; 11:252-256.

17. Spengler D, Bigos S, Martin N, et al: Back injuries in industry: A retrospective study 1. Overview and cost analysis. *Spine* 1986: 11:241-245.

18. Gatchel R, Mayer T, Capra P, et al: Quantification of lumbar function part 6: The use of psychological measures in guiding functional restoration. *Spine* 1986; 11:36-42.

19. Gatchel R, Mayer T, Capra P, et al: Millon Behavioral Health Inventory: Its utility in patients with low back pain. *Arch Phy Med Rehabil* 1986; 67:878-882.

20. Ward N: 1986 Volvo Award in Clinical Sciences: Tricyclic antidepressants for chronic low-back pain: Mechanisms of action and predictors of response. *Spine* 1986; 11:661-665.

21. Mayer T, Gatchel R, Kishino N, et al: A prospective short-term study of chronic low back pain patients utilizing novel objective functional measurement. *Pain* 1986; 25:53-68.

22. Mayer T, Gatchel R, Kishino N, et al: Objective assessment of spine function following industrial injury: A prospective study with comparison group and one-year follow-up; Volvo Award in Clinical Sciences. *Spine* 1985; 10:482-493.

# Part VIII

# Psychological Variables and Issues

After physical evaluation and treatment of the injured worker, restoration of health and return to work are expected outcomes. Although this occurs in most cases, a portion of injured workers do not respond well to medical treatment and eventually become "chronic" patients. At times, there may be a physical problem that is not diagnosed or not thoroughly treated. Many times, however, psychological factors become an important part of the chronic condition.

The term "malingerer" is often misused in an effort to simplify this problem. However, neither the side that wishes to stigmatize the nonworking employee nor the side that prefers to prove disability wins when emotion-laden labels are used.

The sections in Part VIII address the role of psychological variables in restoration of health and return to work. They provide a comprehensive approach to patients who are blocked from functional recovery by psychological problems. Improved knowledge of the syndromes and effective intervention may reduce the alienation caused by subjective judging, which can take place when return-to-work delays are not clearly a result of physical causes.

## Section 17  Psychological Barriers to Recovery

The types of problems seen in situations of delayed recovery are defined and described. This section provides an insight into the variety of premorbid personality characteristics that effect emotions and behavior after a work injury. Clinically relevant examples are offered to enhance understanding and management of these psychological conditions.

### Section 18   Symptom Magnification Syndrome

The symptom magnifier is described and an objective method of working toward functional recovery is outlined. The separate types of symptom magnification are explained and clear examples given. This section documents an original method of classifying psychological attributes in order to further the return-to-work process.

### Section 19   Chronic Pain Management

An overview and model of a successful multidisciplinary pain management program are presented. This approach highlights the specific contribution of team members by discipline and by goals. A sample format is described and case studies are discussed.

## Objectives for Chapter 6

1. List the reasons for using instructional plans in nutrition counseling.
2. List the three steps of instructional planning.
3. Describe research on the advantages of objectives in learning and achievement.

# Section 17

# Psychological Barriers to Recovery

*Lynne E. Killian*

PSYCHOLOGICAL ASPECTS OF DELAYED RECOVERY

Clients who have suffered an industrial accident and who are receiving compensation for the injury may have a disproportionate disability and delayed recovery because of reinforcers provided by the accident. Delayed recovery refers to a group of conditions that prevent injured workers from regaining their productivity or ability to cope at work or in their personal lives. To understand the cause of delayed recovery complications it is necessary to understand behavioral and psychological factors that existed before injury, as well as those that arose at the time of injury and after the injury and treatment process.[1]

**Pre-existing Factors**

The psychological and behavioral factors in existence before the injury can be understood best by evaluating and understanding the client's personality before injury. It is important to note pre-existing adjustments and behavior patterns at work and at home. Injuries and delayed recovery complications may occur in people who have a specific psychological set and emotional status. Consider the following seven factors that can impact on injury and recovery.*

1. *Depression* may predispose a person to accidents. When a person is depressed, psychomotor retardation, inattention, poor visual coordi-

---

*The following factors were adapted from *Compensation in Psychiatric Disability and Rehabilitation* (pp 43–45) by JJ Leedy with permission of Charles C Thomas, Publisher, Springfield, Illinois, © 1971.

nation, and low self-esteem all contribute to accident proneness and poor response to treatment. Supervisors and medical personnel should be alert for signs of depression in workers. Often the complaint of fatigue, hypochondriacal preoccupation, or the appearance of indecisiveness in a usually confident worker may be the overt feature of depression. The depressed client rarely takes proper care of himself or herself and is not committed to a treatment program.

2. *Hysterical personality* is a term used to describe people who are very suggestible and whose inner conflicts are often resolved by developing conversion reactions. An accidental injury can trigger these reactions. Conversion reactions are nonanatomical disturbances of sensory or motor function. The symptoms become the focus and provide a means to avoid dealing with an unconscious or inner conflict. The underlying conflicts often stem from feelings of inadequacy, the desire to be cared for, or the desire to receive special attention or reinforcement from family members. When an accidental injury occurs there is an opportunity to take a passive or sick role and to solve the underlying conflicts without becoming aware of and feeling responsible for them. Complications occur with prolonged and unnecessary diagnostic and treatment procedures that focus on the symptoms.

3. *Dependent, immature* people are likely to surrender to disability as soon as they are injured. These people have not developed a healthy capacity to feel responsible for themselves or to accept work responsibilities. They have a low capacity for work and feel relief when work periods come to an end. They may have a history of job changes with decreasing lengths of time spent on each successive job. They may be quite happy to accept a low level of assured income from disability payments.

4. A more complex pre-existing personality is the *pseudo-self-sufficient* person who needs to maintain self-esteem through work, but this need is in sharp conflict with the desire to be taken care of and to escape responsibility. Although these clients may take longer than expected to recover from injury, especially when treatment requires a prolonged period of dependency, they usually can be remotivated provided the rehabilitation team is aware of the conflict. The longer the period of enforced passivity, however, the more difficult the task of rehabilitation.

5. The *aging* worker with a pre-existing personality disorder may find that his or her injury provides a "way out" of an intolerable work situation. This may be likely if he or she is employed in heavy construction work or an unskilled job with a decreased capacity for work, or if the

person is haunted by the fear of unemployment or unemployability. Workers' compensation following an injury provides a type of early retirement. Failure to understand this often leads to frustrating efforts to treat the client who has little motivation to get well. A new, less taxing, and more satisfying occupation should be the focus of rehabilitative efforts.
6. The *alcoholic* worker is a prime candidate for accidental injury. Even the carefully controlled alcoholic who drinks only after hours is likely to have his or her vigilance and coordination impaired by frequent hangovers.
7. Finally, the *sociopath* and other exploitive people are those who live by their wits and are prone to complications. They like action and tend to work in jobs where injuries are more likely to occur. They have a tendency to manipulate others and prefer work situations with low responsibility. An injury can be used as a means to guaranteed subsistence. When malingering occurs, it is in this group, although not all people in this group are malingerers.[2]

**Traumatic Factors**

Factors that play a role in delayed recovery occurring at the time of the injury are called traumatic factors. Traumatic is a term referring to psychological distress. The injured worker whose disability does not seem to be alleviated often is suffering from "traumatic neurosis."[3]

Traumatic neurosis refers to clients who have developed neuroses following a specific, demonstrable event such as accidental injury. After this event, there is a typical pattern of emotional responses marked by anxiety. The traumatic event or injury creates a state of disequilibrium at the unconscious level that is marked by overwhelming fear within the accident victim. Whether this fear leads to traumatic neurosis depends in part on the person, on the treatment received, and on the person's degree of stress tolerance. If this disruption leads to anxiety there will be psychophysiological responses to the trauma. "Such symptoms may include: startle reactions, fearfulness, repetitive dreams or nightmares reliving the traumatic event, sweating, palpitation, headaches, dizziness, irritability, fatigue, and decreased appetite or sexual drive."[4] Underlying these symptoms is a great deal of anxiety that requires some form of discharge. The client feels overwhelmed, and may withdraw his or her attention from the outside world and become preoccupied with symptoms and emotions. Marked feelings of dependence develop because the client loses some of the ability to function independently and to master or control the environment. The client's friends and family members, seeing the changes in their loved one, may encourage the person to pursue compensation or litigation. Between peer pressure and the client's own anxiety, professionals involved in assisting the worker's return to work are often faced with opposing forces, and efforts may result in little or no improvement. Careful attention to the emotional climate of the injured worker is the way to prevent traumatic neurosis. Education for those involved in the worker's case and the client and family will assist in early recognition and treatment of traumatic anxiety.

**Post-Traumatic Factors**

Somewhere between neurotic anxiety and a frank case of malingering is a classification that describes clients who claim to have symptoms after injury but for whom the traumatic event does not appear sufficient to have caused the symptoms. This subtype of neurosis is often called compensation or "secondary gain" neurosis.[5]

Secondary gain neurosis is recognized as an unconscious phenomenon, as in the anxiety-type neurosis described above. It explains those factors that occur after the injury that contribute to the maintenance of symptoms and inhibition in work performance that may have been avoided if only the pre-existing and traumatic factors were in operation. In other words, secondary gain is an advantage occurring after the accident that plays a part in creating or maintaining the neurosis.

Secondary gain often may be confused with malingering. Many may think that the client is play acting solely to attain monetary gain. If this were true, the client would be considered a malingerer. Secondary gain differs from traumatic anxiety in that the person who exhibits the former seeks gratifications that are dependent (a livelihood provided by others) or retaliatory (against an organization or person believed to be exploitive or injurious to the client).[6]

Secondary gain can take many forms. A person may be thoroughly bored with his or her job or unable to handle the physical stress of the job. An accident would give relief, or secondary gain from these problems. A person may be angry at his or her employer or spouse, or at an insurance company and desire revenge. An accident can provide an avenue for revenge; this is an example of secondary gain. A person may have deep-seated emotional problems that he or she is unable to cope with any longer and an accident can change the stresses in his or her life and give relief. This too would be secondary gain. Thus, secondary gain can be much more than financial gain from the accident.[4]

Treatment and prevention of this psychological complication occurring after the injury involves prompt settlement of litigation or insurance to cover costs only, avoidance of temptation to continue receiving unearned income, and a strong emphasis on rehabilitation rather than on financial compensation.

As a final classification there is the true malingerer. Malingering refers to the conscious and intentional simulation of symptoms—the deliberate imitation of disability for gain. The key words are conscious and intentional. In cases of secondary gain, traumatic neurosis, and pre-existing personality states, the motivating forces are at an unconscious level; for a malingerer the motivating forces are at the conscious level.

True malingerers are rare. Although some patients are malingerers, the vast majority who are injured and fail to improve after a reasonable length of time are not malingering. In the true malingerer, the most frequent personality make-up is that of the antisocial personality.[2] The person is likely to have a poor work attendance and performance record. If malingering is suspected, the patient should be examined carefully and told that his or her physical examination is normal. The key element of the treatment is to get the patient back to work, at all costs, so that the cycle of disability does not begin.[7]

In summary, because the person who has suffered an industrial injury is especially susceptible to psychological complications, it is unwise to assume that patients who have failed to recover in the expected period of time are malingering. Only a small fraction of these patients are malingerers. Although most often pre-existing psychological states interact with the recovery process, the accident process may cause psychological distress and anxiety, and secondary gain may feature a role in retaliatory or dependent gratification. Careful attention to the emotional climate of the injured worker and education of all people involved with the case will help prevent and foster early recognition and treatment of these psychological complications.

## ROLE OF THE PSYCHOLOGIST IN INDUSTRIAL MEDICINE PROGRAMS

The scope of psychological complications and their role in delayed recovery is complex. Quite often, the physician and treating therapist(s) are alerted to possible delayed recovery. This usually takes the form of lack of progress toward rehabilitation goals, decreased motivation, overt signs of depression, anger or anxiety, and noncompliance in the treatment program. At this point it is important for the treatment team to refer the patient to a psychologist for evaluation or consultation and possibly treatment.

## Evaluation

A psychological evaluation involves assessing the person's emotional status and identifying those psychological complications and contributing factors specific to pre-existing personality states, traumatic neurosis, secondary gain, and malingering. Other important factors to be assessed include changes in life style before injury; behavior and responsibilities at home and at work; family and marital relationships; and the client's perception of his or her employer, co-workers, and supervisor.

In addition, the evaluation must identify contributing factors in the area of emotional, behavioral, cognitive, and social domains. These contributing factors may directly or indirectly play a role in the maintenance of the physical injury and associated symptoms, and complicate recovery and effective management of the injury.

Assessing for emotional factors includes a review of the client's affective status and psychosomatic complaints.

Assessing the behavioral domain includes identification of those behaviors in which the client engages when experiencing pain, as well as ways of reducing the pain. The behavioral assessment identifies the reinforcers provided for engaging in these behaviors. In addition, a review should be made to identify maladaptive habits such as dependency on medications, alcohol, and nicotine.

The cognitive assessment includes the client's self-perception of pain, degree of emotional and intellectual insight, and whether his or her affective status is interfering with his or her cognitive abilities (ie, difficulties with memory and concentration).

Finally, the social assessment identifies those life style changes that have occurred since the injury. A comprehensive evaluation will provide valuable information about the client that will assist in the development of an effective treatment program.

## Consultation

A psychological consultation with the treatment team will provide the necessary education on psychological complications in delayed recovery and suggested treatment techniques to use in the client's rehabilitation program. Therapists need to be aware of how to react to these complications. For example, in the case of a traumatic neurotic reaction, it is important to listen to the client's concerns and take those concerns seriously. On the other hand, with a person seeking secondary gain, the therapist should avoid overplaying the sympathetic role, which may prolong emphasis on the injury or reinforce pain behaviors. Therapists should be aware of how an empathetic role with the "wrong" client can lead to condoning or nurturing of symptoms. It is important to remember that the client, particularly in the areas of traumatic neurosis and secondary gain, needs some successes and needs to become actively involved in his or her rehabilitation program. Listed below are some case examples of common "problem clients" whom therapists find difficult to treat.

### Case Example #1

*Problem:* My client is so focused on his pain that he is reluctant to do any activity in work hardening. He is physically stable but won't work up to his maximum potential. He appears uncooperative and frequently cancels treatment sessions.
*Suggestion:* Initially, document all behaviors and subjective responses the client engages that affect his work hardening goal attainment. Then confront the client and review all documented material with him and explain how this is affecting his progress. Although it is important to acknowledge that the client does experience pain, it also is necessary to reduce any reinforcers the client may benefit from when engaging in pain behaviors. Such examples in work hardening may include breaks during treatment; change or reduction of activity in the treatment program; and condoning, sympathy, and attention from the therapist. In addition, provide feedback in graphic form outlining the goals the client has attained, the goals yet to be achieved, and the estimated duration and length of treatment program. Finally, have the client sign a treatment compliance contract indicating that he understands the program's goals and will be discharged if he misses three consecutive unplanned treatment sessions.

### Case Example #2

*Problem:* My client has developed a good relationship with me and trusts me. Yet, she feels I should be more sympathetic and understanding to her physical weaknesses. She doesn't feel I should push her hard on her exercise program and she seems withdrawn. How can I progress her without alienating her?

*Suggestion:* Refer this client to the psychologist to evaluate for possible depression. If the depression is primary it will most likely need to be treated before you can expect any improvement in her work hardening program. Maintain your objectivity and emphasis on treatment goals while providing supportive communication and performance feedback. The therapist and psychologist can work together as a team to facilitate patient success.

**Case Example #3**

*Problem:* I think my client is not as disabled as he wants me to think he is. His performance in work hardening is inconsistent and does not seem to be based on any physical problems. How can I get him to perform at his true level? He appears to have difficulties with concentrating and his memory isn't very good. I don't want him to think he is fooling me, but I hate to call him an intentional malingerer.

*Suggestion:* Document all objective and subjective discrepancies in performance and compare this information with the medical evaluation of his injury. Notify the client's referring physician or case manager and refer for a psychological consultation or evaluation. It is appropriate to make a referral to a psychologist not only when personality and emotional problems are suspected, but also when there is questionable performance in the cognitive-behavioral areas, such as following instructions, memory and concentration, and social interaction. In addition, a psychologist can start to lay the necessary groundwork for vocational rehabilitation by providing intelligence and vocational interest testing when warranted.

**Treatment**

Psychological treatment may be indicated particularly with identified delayed recovery complications. In some instances, there may be reluctance on the part of the insurer to assume responsibility for these additional costs. The insurer may not acknowledge the role of psychological complications and target the client as a malingerer. Or, the insurer may recognize the client's psychological distress but view it as "premorbid." Such a conflict fosters medicolegal and ethical concerns. Education of all those involved with the worker's case and careful attention to the psychological effect of the accident on the worker will assist in minimizing such conflict and will foster a cost-effective and successful rehabilitation program.

The psychologist in an industrial medicine program must provide education and support for injured workers and their families to facilitate insight into their psychological complications and reentry into the work role. This can be implemented best by using a group process whereby a psychologist provides a lecture during each group session. The educational component should be followed by a supportive discussion with the group members so they will personally address those issues related to the lecture. The educational component should include lectures on the psychological benefits of the work role and developmental changes after injury, motivation and self-defeating emotions, delayed recovery, pain and stress management, and avoidance conflicts.

Clients need to recognize the importance of the work role for the psychological benefits and social reinforcement it provides. Such benefits include social continuity, the sense of mastery and control in the environment, status and peer recognition, and the opportunity for new learning activity. Emotional disruption, and its effect on the individual and on the family and its role in symptom magnification, should be addressed. In addition, clients must learn how emotions such as anger, anxiety, and frustration play a role in psychosomatic complaints and delayed recovery.

The relationship between self-defeating emotions and motivation should be reviewed. In addition, techniques from cognitive therapy, the importance of goal setting, and communication skills should be emphasized. Group members need to be aware of the pre-existing personality positions that complicate recovery and need to be encouraged to address these as they may relate to them.

The "accident process" and the factors that occur before, during, and after the injury that may lead to a delayed recovery should be discussed. Post-traumatic neurosis, secondary gain, and malingering are such outcomes of delayed recovery and need to be explained to the clients. The importance of early rehabilitation activity, goal setting, and supportive communication should be emphasized.

Pain and its relationship with stress in the pain-tension cycle is another area for client education. Clients should learn how to practice several pain management techniques that reduce and relieve stress. Progressive muscle relaxation and visual imagery exercises should be provided.

Finally, the role of anxiety with regard to reinjury and returning to work may foster an avoidance conflict that is important to overcome. Preventive measures that may alleviate anxiety and facilitate a successful return to work should be provided to injured clients. In addition, the potential attitudes and behaviors of co-workers, employers, and supervisors should be discussed. Techniques and suggestions should be provided for the employee who has to return to a work setting with co-workers that have a negative attitude. A joint conference with the employee, employer, supervisor, rehabilitation consultant, physician, and work hardening team is recommended. Work adaptations, physical limitations, and other changes should be clearly communicated at the conference to reduce anxiety and to foster a successful return to work.

In summary, education and early recognition of the psychological complications specific to industrial injuries are the keys to prevention of delayed recovery. Careful attention to the psychological climate of the injured worker and education of physicians, rehabilitation counselors, therapists, insurance adjusters, employers, and the patient and family will prevent some of the psychological complications that occur during and after an industrial accident. Early rehabilitation, treatment, and education should be the guiding concepts in the handling of an injured worker's claim.

**REFERENCES**

1. Ross WD: How to get a neurotic worker back on the job successfully. *Occup Health Saf* 1977; 46:201-23.

2. Enelow AJ: Malingering and delayed recovery from injury, in Leedy JJ (ed): *Compensation in Psychiatric Disability and Rehabilitation*. Springfield, Ill, Charles C Thomas Publisher, 1971, pp 42-46.

3. Millen FJ: Post-traumatic neurosis in industry. *Ind Med Surg* 1966; 35:929-935.

4. Keiser L: The traumatic neurosis, in Leedy JJ (ed): *Compensation in Psychiatric Disability and Rehabilitation*. Springfield, Ill, Charles C Thomas Publisher, 1971, pp 272-286.

5. Woodyard JE: Diagnosis and prognosis in compensation claims. *Ann R Coll Surg Engl* 1982; 64:191-193.

6. Ross WD: Differentiating compensation factors from traumatic factors, in Leedy JJ (ed): *Compensation in Psychiatric Disability and Rehabilitation*. Springfield, Ill, Charles C Thomas Publisher, 1971, pp 31-41.

7. Derebery VJ, Tullis WH: Delayed recovery on the patient with a work compensable injury. *J Occup Med* 1983; 25:829-835.

# Section 18

# Symptom Magnification Syndrome

*Leonard N. Matheson*

Symptom magnification syndrome is a new concept in the field of health care. As a new concept, much of the work is yet to be done. However, it appears that sufficient progress has been made in the study of the syndrome to present an introduction that will prepare the reader to understand and identify its existence as a significant component of most professional rehabilitation practices.

## BACKGROUND

As with any new concept, the symptom magnification syndrome did not begin with a fully developed definition but has evolved with experience. Symptom magnification syndrome was originally described by Matheson as "a constellation of reports or displays of symptoms which functions to control the environment to such a degree that the symptom behavior is maladaptive."[1]

As an understanding of the symptom magnification syndrome developed, further elaboration became necessary. In an invited address to the National Rehabilitation Association, the author expanded the original description to include a focus on the social learning aspect of the behavior, so that symptom magnification syndrome was seen as "self-destructive behavior in which reports or displays of real or imaginary symptoms are socially reinforced."[2] In a later paper presented to that same organization, the author pointed out that these maladaptive behaviors are not necessarily conscious, so that symptom magnification syndrome "may be either an unconscious or a conscious learned behavioral response pattern in which symptom behaviors control the life circumstances of the sufferer to such a degree that health care efforts are impeded."[3]

These descriptions have been in use in the author's clinical practice. Although the concept of symptom magnification syndrome may evolve even further, no longer is it necessary to postpone the presentation of a formal definition. The author offers a current, consolidated definition of symptom magnification syndrome as a conscious or unconscious self-destructive socially reinforced behavioral response pattern consisting of reports or displays of symptoms that function to control the life circumstances of the sufferer. This definition encompasses all of the important aspects of the syndrome; that is,

- It may be conscious *or* unconscious.
- It is self-destructive.
- It is a *pattern of behavior* that is learned and is maintained through social reinforcement.
- It is composed of reports *and/or* displays of symptoms.
- Its effect is to control the life circumstances of the sufferer.

## DEVELOPMENTAL ORIGINS

In this society, there is a tacit agreement that people will assist each other when trouble occurs. Reports and displays of symptoms are one indication of trouble. One of the primary functions of symptoms is to control the environment. The alerting, warning, and mobilizing functions of symptoms are powerful personal and social phenomena and are learned early in life, certainly before language and probably as one of the first cognitive skills. An infant learns that the "reports or displays" of symptoms can control the external environment, principally the parents. With maturity, the person learns to internalize direct control over the environment and, consequently, to rely less on symptoms as a means of control.

What happens when a person experiences the rapid, unexpected onset of pathology that leads to a chronic disability? Initially, the alerting and warning effects of the reports and displays of symptoms are of crucial importance. Great attention is paid to them to effect an accurate diagnosis of the pathology and to develop a treatment plan. As the pathology is diagnosed and whenever subsequent impairment becomes stable, the reports and displays of symptoms become less important. Without an emergent crisis, the alerting effect of the symptoms is usually greatly diminished.

With a responsible adult patient who has learned to care for his or her impairment so as to avoid exacerbation, the mobilizing effect of the symptoms is greatly diminished. Because the chronic problem is clearly understood and its symptoms are anticipated, the mobilization of resources can be planned and easily controlled by the patient. All that the patient needs to do is "pop a nitro" for his or her angina or chew a sugar cube in response to an insulin reaction. These are both examples of symptoms that are indications of life-threatening problems that often are treated by patients with excellent effect. In successful rehabilitation, the responsibility for symptoms and whatever control over them is possible is relinquished to the patient whenever the patient is competent to handle that responsibility.

The person who experiences rapid, unexpected onset of pathology that leads to a chronic disability may not successfully take responsibility for his or her symptoms and, in fact, may demonstrate a symptom's function to control life circumstances because he or she perceived the emergent situation and its aftermath as uncontrollable. The person's response to this uncontrollable situation may make him or her susceptible to the symptom magnification syndrome.

How does the susceptible patient become caught in the syndrome? Unfortunately, it is usually through the well-guided efforts of the same professionals who are attempting to achieve his or her rehabilitation.

Health care professionals who are trained to respond to reports and displays of symptoms in a helpful, caring manner use symptoms as an important part of the relationship with the people whom they serve. In fact, the symptom display behaviors are so important that they are reinforced by health care professionals who rely on accurate, precise perception of these behaviors to diagnose and to treat.

As long as the reports and displays of symptoms are simply part of communication and are differentially reinforced as simply communication, symptom magnification syndrome will not result. However, symptom magnification syndrome becomes an iatrogenic problem when the reports and displays of symptoms begin to control the relationship. The health care professional becomes responsible for precipitation of the syndrome in a susceptible patient.

What distinguishes a susceptible patient? Experience with a large clinical population suggests that the concept of "learned helplessness"[4] may hold important clues to the development of susceptibility. The following case example helps to illustrate the issues.

> Calvin R. is a 49-year-old male shipyard welder. Working deep in the hull of a ship one day, he began to experience rapid onset of disabling angina. He laid down immediately and weakly called for help. The noise of the ship was such that his calls went unanswered.
>
> Four hours later, well after his shift had ended, Calvin's wife called the shipyard inquiring as to his whereabouts. Calvin's supervisor noted that

his car was still in the parking lot and that his toolbox had not been returned. The supervisor returned to the ship and found Calvin lying down, semiconscious, and in critical condition. He was taken to the coronary care unit at a nearby hospital and survived.

Six weeks later, the author had the opportunity to begin working with Calvin. Coronary angiography had demonstrated a very small segment of blockage distally in one coronary artery, immediately above a small area of infarct. Exercise stress testing showed a normal cardiac response. However, as Mr. R. prepared to return to work, his angina returned and became disabling. He had an atypical response to nitroglycerin and an inconsistent response to activity. In areas other than those that focused on a return to work, he was not apparently disabled.

Mr. R.'s predominant emotional expression was anxiety. Treatment began with systematic desensitization assisted with the use of a mixture of carbon dioxide and oxygen to bring about profound relaxation. In addition, multichannel biofeedback was used to assist Mr. R. to gain a sense of mastery over his symptoms. Counseling sessions with his wife were begun to develop an ally at home that would assist him in talking about the fear that he had about a recurrence of the episode that had immediately preceded his hospitalization. Mr. R. readily admitted that he could think of no more frightening experience than that episode.

A visit was made to the shipyard and ship with Mr. R. A small bottle of oxygen and oxygen mask were made available to Mr. R. in his toolbox, along with a very loud alarm should he have similar difficulty in the future. All of these were designed to provide Mr. R. with a sense of control over the circumstance to which he anticipated returning.

Mr. R. successfully returned to work and experienced no additional symptoms.

Helplessness is the psychological state that results when events are uncontrollable. The newly disabled patient learns helplessness when he or she experiences "response-outcome independence"—that is, when the patient feels that no matter what he or she does (or does not do) events proceed without his or her apparent control. The experience is perceived as uncontrollable.

The traumatic onset of an injury, which is likely to result in chronic impairment and loss of job, is an event that causes a heightened state of emotionality. If this is a work-related injury treated in the context of the workers' compensation system, an important focus of loss for the person is the loss of control over his or her life's circumstances. The person finds himself or herself enmeshed in a societal system that is rigorously regimented and, he or she believes, is unfair. Quite important is the fact that the injured worker experiences a loss of control over his or her life at the same time that he or she experiences the single most important blow to the sense of competence (ie, loss of function) that the person may experience in a lifetime. The loss of real control over one's life circumstances coupled with the loss of sense of competence easily can be seen to lead to a profound sense of helplessness.

The expectation that this outcome is uncontrollable and independent of whatever response may be initiated by the injured worker produces several outcomes:

1. The motivation to attempt to control the outcome is reduced. Helplessness saps the motivation to initiate voluntary responses that control other events. Voluntary responding requires the expectation that the response will succeed. The experience of uncontrollability leads people to the belief that responses will not succeed. Hence, responses that are used to attempt to control the person's life experience are not attempted. This becomes a self-fulfilling prophecy. It appears to the observer as a motivational deficit.
2. Learning that the response can control the outcome is disrupted. An expectancy is developed that future efforts will be futile. Once a person has had experience with uncontrollability, he or she has difficulty learning that the response has succeeded, even when it is actually successful.

   Uncontrollability distorts the perception of control. This is called a "negative cognitive set." This negative cognitive set is such that people believe that success and failure are independent of their own action and therefore they have difficulty learning that subsequent (actually effective) responses work.
3. Early in the process of adjustment to disability, the person is anxious and fearful. Although this emotional state will dissipate in time, it may become far more serious and lead to a major depression. As the disability becomes clear, anxiety and fear diminish and the person begins to adjust. It is at this point, if the person believes that he or she is helpless and unable to control the consequences of the injury, that a disability-induced reactive depression can occur.

The value of the symptom magnification syndrome in rehabilitation centers on the concept's ability to assist those in rehabilitation to work effectively with the patient so that the patient can drop the use of reports and displays of symptoms as a means to control the environment and cope with helplessness.

262  WORK INJURY

```
┌─────────────────────┐
│      SYMPTOM        │
│   MAGNIFICATION     │
│   DECISION TREE     │
└──────────┬──────────┘
           │
┌──────────┴──────────┐
│    Symptoms are     │
│   Non-Negotiative   │
└──────────┬──────────┘
           │
┌──────────┴──────────┐
│  Symptoms Result in │
│ Environmental Control│
└──────────┬──────────┘
           │
┌──────────┴──────────┐
│  Symptoms Magnify   │
│ Functional Limitations│
└──────────┬──────────┘
           │
┌──────────┴──────────┐
│     Confirm SMS     │
└─────────────────────┘
```

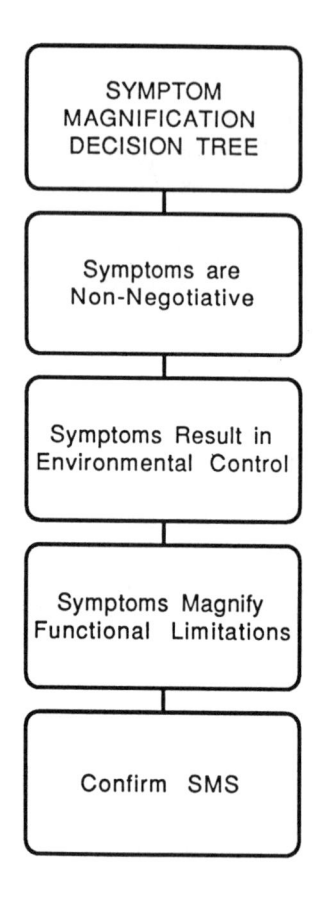

**Figure 18-1** Symptom magnification decision tree. *Source:* Reprinted from *Symptom Magnification Casebook* (Chapter 2, p 7) by L Matheson with permission of Employment and Rehabilitation Institute of California, © 1987.

## IDENTIFICATION

A diagnostic decision tree can be used to confirm the presence of symptom magnification syndrome. (See Figure 18-1.) The focus of the identification process is, first, on the symptoms' "nonnegotiability"; second, on the *effect* of the behavior pattern; and third, on the degree of magnification of the expected functional limitations.

The order of importance of factors in this decision is in distinction to a primary consideration about whether or not the symptoms are florid and widespread. It is *not* correct to say simply that the degree of symptomatology is predictive of the presence of the syndrome. The concern about symptoms is whether or not they are "negotiable," that they can be manipulated through control of activities. If the symptoms can be made worse or better through the sufferer's participation in certain activities, the symptoms are controllable by the sufferer. The symptoms are not "in control" of the sufferer and can be seen to be the responsibility of the sufferer. This responsibility is incompatible with the definition of the syndrome presented earlier in that the symptoms cannot be both the responsibility of the suffer *and* the "function to control the life circumstances of the sufferer."

Although it is premature to take a definite stand on the prevalence of the syndrome in society, its prevalence in an industrial rehabilitation clinical setting can be studied. With a total of 377 industrial rehabilitation cases on which complete data were collected, the prevalence is 24% (91 total identified cases). Figure 18-2 describes this sample, further delineated by gender.

Prevalence of Symptom Magnification Syndrome
(Clinic Total = 377)

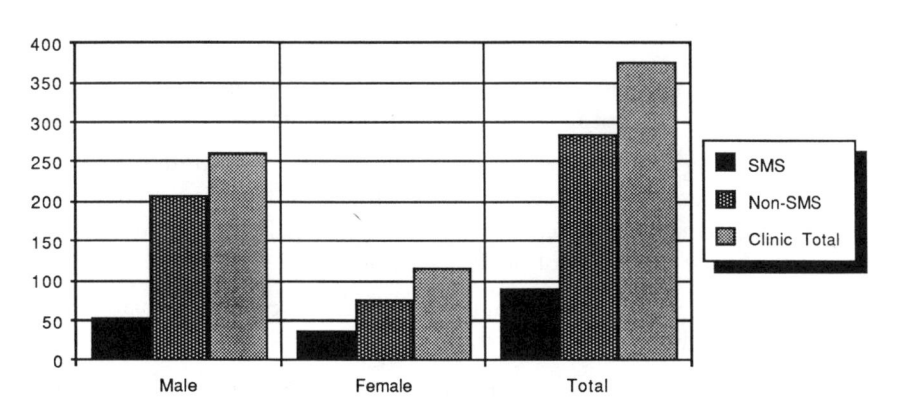

**Figure 18-2** Prevalence of symptom magnification syndrome. This caseload is composed of people who sustained industrial injuries and participated in a rehabilitation program that was underwritten by the employer or workers' compensation insurance carrier. Subjects in this sample are typically more than 2 years postinjury, possess less than a high school education, and do not have a job available to which to return. These early data suggest a gender bias, in that with only 31% of the caseload women represent 42% of the syndrome cases. *Source:* Reprinted from *Symptom Magnification Casebook* (Chapter 1, p 13) by L Matheson with permission of Employment and Rehabilitation Institute of California, © 1987.

## Types of Symptom Magnifiers

Three primary types of symptom magnification syndrome sufferers have been identified. Each of the symptom magnification syndrome types has an analog in the psychiatric literature. However, because this disorder is considered in terms of a behavior pattern rather than in terms of a psychiatric syndrome that requires knowledge of intrapsychic function unavailable to the nonpsychotherapeutic clinician, effective intervention can be offered within the context of vocational rehabilitation.

### *Type 1 Symptom Magnifier*

The type 1 symptom magnifier (called the "refugee") discovers that the symptom behavior provides an escape from an apparently unresolvable conflict or life situation. The refugee is looking back over his or her shoulder while *escaping* from a difficult life situation and has little future orientation. Goals are very difficult for this person to develop or embrace because there is so little future orientation. The refugee has a psychosocial history that includes a placement in the family or the community network that the person perceives as being integral and irreplaceable to that family or community. This person perceives the availability of effective role backup as nonexistent. Behaviorally, the refugee

- is in and out of the worker role; often is able to work part time on an occasional basis but never in a sustained manner
- is unwilling to seek support, especially under fire, (although he or she may appear to be "helpless"); is willing to grit his or her teeth and to endure a conflict that actually appears to him or her to be unresolvable without the assistance of outside support
- appears as a martyr in relation to the symptoms (eg, may verbalize "The pain is terrible but I'll make it through somehow")
- will often involve professional caregivers in "yes-but" interchanges as they attempt to provide assistance or guidance; symptoms provide an escape from the unresolvable conflict.

The analog to the refugee in the psychiatric literature is the person who suffers from a somatoform disorder. According to the *Diagnostic and Statistical Manual of Mental Disorders* (DSM-III),[5] (p 19) this is a group of disorders in which symptoms suggest a physical disorder for which there are no findings or a known physiological mechanism, and for which there is evidence that the symptoms are linked to psychological factors or conflicts. According to DSM-III, this group of disorders includes the following:

- *Somatization Disorder* (DSM-III #300.81)—Characterized by recurrent, multiple somatic complaints of several years' duration. It usually begins before the age of 30 years and has a chronic, fluctuating course. Complaints are usually dramatic but vague. Anxiety and depression, including suicidal ideation and attempts, are usually found in tandem with somatization disorder. Substance use disorders (with prescribed medications) are frequently found in the somatization disordered person.
- *Conversion Disorder* (DMS-III #300.11)—Also known as Hysterical Neurosis—Conversion Type. The conversion-disordered person experiences an actual loss of or alteration of physical function that is an expression of a psychological conflict. This loss is *not* under voluntary control, but cannot be explained by any known pathophysiological mechanism. The conversion-disordered person achieves primary gain by avoiding an unpleasant activity or responsibility, or by getting support from the environment that otherwise would not be available.
- *Psychogenic Pain Disorder* (DSM-III #307.80)—The constant complaint of pain without physical findings *and* with evidence of the probable causative role of psychological factors. With this disorder, a psychological conflict is tied to the occurrence of the pain. Additionally, as with the conversion disorder, the pain may "subside" to allow the person to participate in some activities while avoiding other activities.
- *Hypochondriasis* (DSM-III #300.70)—Also known as Hypochondriacal Neurosis. This problem is characterized by the exaggeration of concern about physical signs, taking essentially normal aches and pains as abnormal and indicative of a disease process. This person is preoccupied with the idea that he or she has a serious disease that persists despite medical reassurance.

*Type 2 Symptom Magnifier*

The type 2 symptom magnifier (called the "gameplayer") discovers that symptoms provide an opportunity for positive gain. The gameplayer is an opportunist who appears to be in the "daydreaming" stage of career development and has a history of strong goal orientation with goals that are few in number, very challenging with high visibility, and impoverished or too simply defined. Goals such as "to own a dump truck and go into the construction business," or "to be a drug counselor" are not backed with any planning, specific knowledge, or attempted action.

Behaviorally, the gameplayer

- has a history of extravagant goal setting with poor goal attainment
- displays individual-relevant symptom behavior
- has great variability in maximum performance levels
- acts impulsively, in heroic disregard of his or her impairment.

Further investigation into the cause of the gameplayer's poor goal attainment will demonstrate that he or she is generally a poor planner. The gameplayer tends to be irresponsible while appearing to be responsible. When things go wrong he or she will say, "It's not my fault; I told her she should have done such and such." In many cases, the gameplayer will have an antisocial personality disorder (according to DSM-III #301.70).[5] He or she actually has few if any true friends or significant others whom he or she can trust and who will rely on him or her. Conversely, he or she often has a large number of acquaintances and a wide social circle.

The gameplayer is similar to the person who traditionally has been considered the "malingerer," which is the most appropriate analog in the psychiatric literature. According to DSM-III the malingerer is presented as a "V-Code," which characterizes "conditions not attributable to a mental disorder that are a focus of treatment or attention."[(p 331-332)] Because it is so often applied inappropriately, the DSM-III definition will be presented in its entirety:

> V65.20 Malingering—The essential feature is the voluntary production and presentation of false or grossly exaggerated physical or psychological symptoms, produced in pursuit of a goal that is obviously recognizable with an understanding of the individual's circumstances rather than of his or her individual psychology. Examples of such obviously understandable goals include: to avoid military conscription or duty, to avoid work, to obtain financial compensation, to evade criminal prosecution, or to obtain drugs. Under some circumstances Malingering may represent adaptive behavior, for example, feigning illness while a captive of the enemy during wartime. A high index of suspicion of Malingering should be aroused if any combination of the following is noted:
>
> 1. medico-legal context of presentation, eg, the person's being referred by his attorney to the physician for examination;

2. marked discrepancy between the person's claimed distress or disability and the objective findings;
3. lack of cooperation with the diagnostic evaluation and prescribed treatment regimen;
4. the presence of Antisocial Personality Disorder.

The differentiation of Malingering from Factitious Disorder depends on the clinician's judgement [sic] as to whether the symptom production is in pursuit of a goal that is obviously recognizable and understandable in the circumstances. Individuals with Factitious Disorders have goals that are not recognizable in light of their specific circumstances but are understandable only in light of their psychology as determined by careful examination. Evidence of an intra-psychic need to maintain the sick role suggests Factitious Disorder. Thus the diagnosis of Factitious Disorder excludes the diagnosis of the act of Malingering.

Malingering is differentiated from Conversion and the other Somatoform Disorders by the voluntary production of symptoms and by the obvious, recognizable goal. The malingering individual is much less likely to present his or her symptoms in the context of emotional conflict, and the symptoms presented are less likely to be "symbolic" of an underlying emotional conflict. Symptom relief in Malingering is not often obtained by suggestion hypnosis, or intravenous barbiturates, as it frequently is in Conversion Disorder.[5]

Symptom magnification is a much more prevalent problem than is malingering. True malingerers represent a very small proportion of the number of people who use symptoms to control their external environments.

The professional who confuses symptom magnification syndrome with malingering demonstrates a lack of knowledge. It is a serious mistake for the health care professional for two important reasons:

1. Symptom magnification syndrome is treatable and should be looked at like any other self-destructive behavior pattern. Symptom magnification syndrome sufferers need to be identified and treated, not "caught" and "turned into the sheriff." Malingering is not treatable because "malingerer" is a pejorative label, completely at odds with a therapeutic relationship. Instead of using terminology that is disparaging or belittling, the symptom magnification syndrome approach allows the discussion of symptom magnification syndrome behaviors with the patient in a constructive manner. The professional can say,

"This is a normal behavior pattern, but it is self-destructive and should not be continued. I can help you to change." In addition, if the professional is incorrect in applying a malingerer label, he or she has placed a damning label on a person who often desperately needs help.
2. Malingering is not a defensible diagnostic label, but rather a medicolegal concept. Health care professionals may use it as a psychiatric label but, as may be seen above, this is incorrect. Malingering connotes a "crime" for which evidence must be gathered. Leave this to the attorneys and private investigators.

From the author's point of view, the concept of malingering does not work in rehabilitation and should be discarded from the rehabilitator's lexicon. In its place, the symptom magnification syndrome will be found to be less problematic and is certainly less stigmatizing. Use the symptom magnification concept because it connotes something that is treatable and may be identified and defended based on behavioral observation.

*Type 3 Symptom Magnifier*

The type 3 symptom magnifier (called the "identified patient") discovers that symptoms ensure survival and maintenance of the patient role. For the identified patient, the patient role eclipses and contains all other possible roles. For example, for a man, roles of father, husband, brother, uncle, friend, and neighbor and other roles as well are frequently perceived to have been lost. Although the identified patient may continue to have the title, he or she is not treated in the customary manner associated with the role.

Behaviorally, the identified patient

- has a few goals that focus on psychological or physical survival—to maintain the patient role and "to get through the week," "to make it to my next disability check," or "to get this prescription filled"; the "grand scheme" of the gameplayer is not present
- acts as if life is to be survived rather than enjoyed
- acts impulsively, in "accidental disregard" of his or her impairment.

The analog to the identified patient in the psychiatric literature is the factitious disorder, which indicates an artificial disorder. Factitious disorders are characterized by physical or psychological symptoms that are produced by the person and are under voluntary control. Differentiation from malingering is based on the finding that there is no apparent goal for the person other than maintenance of the "patient role."

There are two primary types of factitious disorders[5]:

- *Factitious Disorder with Psychological Symptoms* (DSM-III #300.16) —The voluntary production of severe psychological symptoms that suggest a mental disorder is the essential characteristic of this disorder.
- *Chronic Factitious Disorder with Physical Symptoms* (DSM-III #301.51) —The essential feature of this disorder is the presentation of physical symptoms that are not real. The display of symptoms is limited only by the person's medical knowledge. The presentation may be total fabrication or it may be self-inflicted, including inappropriate self-medication and exacerbation of previous injuries.

### Differentiation Among the Types of Symptom Magnifiers

The symptom magnification syndrome decision tree (Figure 18-1, *supra*) can be further delineated (see Figure 18-3) to include the issues that are considered in arriving at an appropriate type classification.

The issues involved in differentiating among the types of symptom magnifiers must be handled in the order they are presented in Figure 18-3. In addition, it is much easier to identify the general symptom magnification syndrome than it is to pinpoint the type. In many cases, the symptom magnification syndrome type may be identified with any degree of assurance only during treatment.

Significant differences are found by gender with men demonstrating a greater than expected frequency of type 2 gameplayers while women demonstrate a greater than expected frequency of type 1 refugees. These data must be considered a preliminary. Further structured study is necessary and is underway.

### EVALUATION

Evaluation of symptom magnification syndrome is undertaken within the context of the decision tree (Figure 18-1, *supra*).

**270**  WORK INJURY

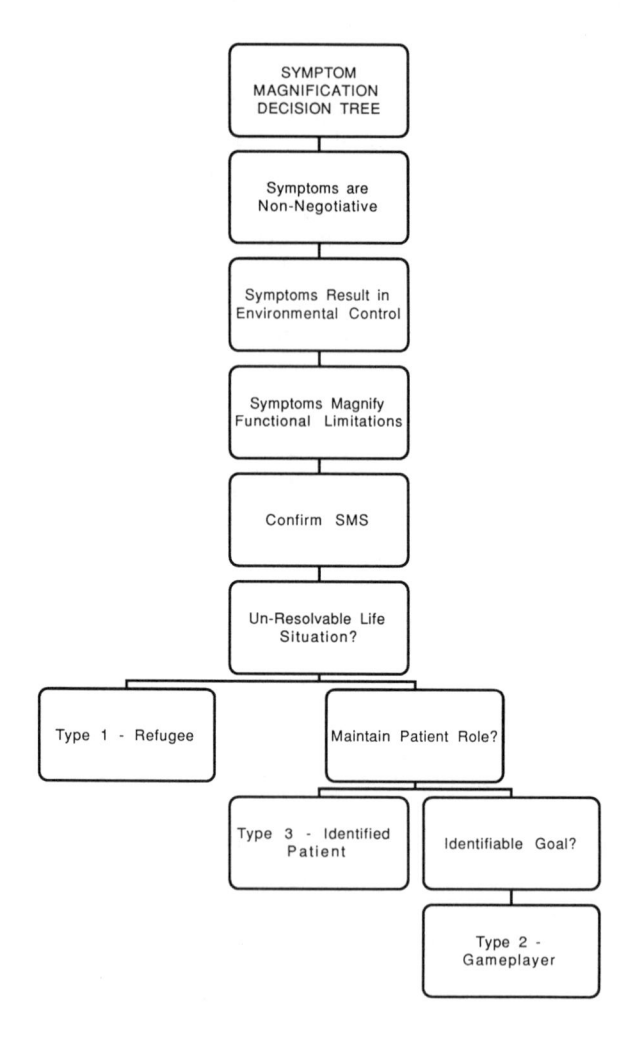

**Figure 18-3** Delineation of the symptom magnification decision tree to differentiate among the types of symptom magnifiers. *Source:* Reprinted from *Symptom Magnification Casebook* (Chapter 2, p 11) by L Matheson with permission of Employment and Rehabilitation Institute of California, © 1987.

## Symptoms Are Non-Negotiative

The symptom magnifier will present his or her symptoms as powerful and immutable, unable to be effectively controlled on a dependable basis. Examination of this issue begins with an interview that includes the following:

- Self-descriptive statements such as those indicating that symptoms control his or her activities without recognition that the relationship is circular.
- A detailed narrative description of the patient's current level of function with a focus on the specificity of the patient's knowledge. Lack of specificity is not terribly diagnostic; most people provide vague descriptions. If there is a lot of specificity, however, the person is not likely to be a symptom magnifier. Vagueness and lack of awareness may indicate that the person is not taking responsibility for developing activities that can control the symptoms.
- Completion of a human figure drawing that graphically depicts the patient's symptoms. This drawing is reviewed with the patient with a focus on the consistency of activity producing the symptoms as opposed to the "magical" occurrence of symptoms without cause. A symptom magnifier often will view symptoms as occurring magically, not related to activity. Inability to respond to the question, "What makes it worse?" may show avoidance of responsibility, because avoiding activities that produce symptoms prevents their exacerbation. Look for statements that externalize the problem, that imply that the person is not responsible for controlling symptoms. Such phrases as, "My back won't let me...It doesn't behave," "Pain won't let me," or "Pain stops me from working" are indicative of external control.

## Symptoms Result in Environmental Control

In the intake interview and in subsequent evaluation tasks the symptom magnifier will depict symptoms as if they maintain control of his or her environment (the physical and social milieu in which he or she operates). Issues such as the following are important to elicit:

1. The historical consequences of the symptom behavior.
    - How have the symptoms previously affected the patient's life?
    - If the patient had a previous injury, did he or she benefit?

2. The present reinforcement structure in which the patient is operating.
   - How will the patient benefit from the maintenance of the symptoms?
   - What is the patient's perception of potential benefit from the maintenance of the symptoms?
3. The community's expectations. Ask the patient:
   - How does your family see you?
   - What does your spouse want from you?
   - Do you think you can provide what your spouse wants from you?
4. The community's support for alternate behavior. If you don't see healthy behaviors leading to return to work being reinforced and fully supported, you can assume that unhealthy behaviors are being reinforced. Ask the patient:
   - Who helps with activities of daily living? with chores?
   - Does this person seem to mind?
   - Does your husband/wife want you to go back to work?
   - Do you get active encouragement to be successful?
5. Adjustments of life style that have been made. Are these retrievable or have they been dismissed?
6. The greatest perceived loss because of the injury. Is this amenable to change? For example, perceived loss of ability to walk is a more problematic response than is lost ability to go shopping. The former may be permanent and harder to put behind. If the greatest perceived loss is something about which nothing can be done, adjustment is more difficult. The response to the question of greatest perceived loss for a dominant-hand amputee suffering from the symptom magnification syndrome might be, "my hand" whereas the response from the person who does not suffer from symptom magnification syndrome might be, "my job" or, "my ability to write." It also is important to inquire after the patient's perception of the permanence of the loss.
7. The patient's stated goals. At this point in the interview, the attitude of the interviewer should be modestly encouraging. Do not offer judgments or comments regarding whether the goals are realistic or unrealistic. Avoid communicating value judgments. A sample question such as, "What do you want most out of rehabilitation?" should elicit a response that can be pursued to deal with these issues:
   - Is the goal medical/curative or functional? "Looking for a cure" could indicate symptom magnification syndrome.
   - Is the goal description explicit or vague? A vague goal such as, "to go back to work" may indicate symptom magnification syndrome.

- Is the degree of goal development rich or simple? Simple goals lacking in detail may indicate symptom magnification syndrome.

Some characteristic responses of the different symptom magnification syndrome types include:
- The type 1 is looking back over his shoulder while escaping from a life situation, so has little future orientation.
- The type 2 acts like he or she is "daydreaming" with goals that are "highfalutin," highly visible, simplistic, and lacking detail.
- The type 3 has a few goals that are impoverished and survival-oriented (eg, "to get through the week").

8. "Work function themes," those unconscious rules by which the patient guides his or her participation in work activities. The *WEST Tool Sort*[6] is used to elicit this issue. The work function themes are unconscious when the evaluee first performs the sort, but become conscious during the debriefing. The patient initially may have (unconsciously) seen himself or herself as being less capable and, with the increased awareness as the debriefing proceeds, may report that he or she is more capable (by shifting cards from the less capable categories to those that indicate greater capability). This initial underestimation of ability is called a "negative transposition." If the patient unconsciously overestimates ability and with increased awareness becomes more realistic (by moving the cards to a lower functional level), this indicates a previous "positive transposition." A preponderance of negative transpositions is indicative of symptom magnification syndrome. If a positive finding results, retest with the *WEST Tool Sort* because it is very hard to fake responses.

**Symptoms Magnify Functional Limitations**

Once the patient's perceptions of his or her functional limitations have been collected through the interview and *WEST Tool Sort*, the consistency of his or her functional performance can be studied. In this area, behavioral observations form the backbone of the evaluation data collection. Reports should be recorded formally in a record designed for that purpose, initialed and dated by the reporter. The key issues to look for are the following:

1. Consistency of symptomatic response to simulated work activities.
   - Consistency of function across various work activities.
   - Dissimilar function demonstrated during similar tasks.

- Behavior variability related to patient's awareness of being watched or "tested"—indicative of symptom magnification syndrome.
2. Inability to develop at least a minimum level of activity control of symptoms. Some level of activity control is always possible *unless* the patient suffers from the symptom magnification syndrome.
3. "Sabotage" of the program by the type 3 identified patient who may do things that cause him or her to be symptomatic so that the patient role is maintained.
4. Externalization of blame by the type 2 gameplayer for not participating or for not succeeding by saying such things as, "The exercises you gave me for transfer of dominance made my back hurt" or, "You insulted me...I'm leaving." Keep a sense of perspective here and record the behavior.

## TREATMENT

Treatment for symptom magnification syndrome focuses on the personal paradigm of the patient. It conceptualizes treatment in terms of a "cognitive revolution" in which the symptom magnifier *views himself or herself*. This is based on the notion that the personal paradigm of symptom magnifiers is a negative cognitive set in which they see themselves in their world as much less capable and much less self-directed than they are and can become. Effective treatment requires identification of the symptom magnifier's view of the world and modification of that view so that it becomes more appropriate for the patient to act directly on the environment than to allow the symptoms to control the environment for him or her.

Treatment for symptom magnification requires engaging the patient's attention and interest to induce the patient to counteract the negative response set and to become involved in constructive activities. Symptom magnifiers often will have become so divorced from participation in social roles that there is no expectation of satisfaction that was previously obtained and that sustained the performance of the roles.

Successful treatment of symptom magnification syndrome involves the use of *graded mastery techniques*. The focus of graded mastery techniques is to identify tasks that the patient can attempt with a reasonable likelihood of successful outcome. The therapist works with the patient to identify these tasks and to set the contingencies in motion for their successful outcome. As the patient demonstrates mastery, the therapist's involvement lessens so that, ultimately, the patient is mastering important tasks on his or her own.

Treatment for symptom magnification syndrome takes place within the context of a work hardening program. Work hardening is defined as

1. A prescriptive productivity development program.
2. A highly structured productivity-oriented treatment program which uses the injured worker's involvement in real or simulated work activities as its principal means of treatment.
3. An individualized, work-oriented treatment process involving the injured worker in simulated or actual work tasks that are structured and graded to progressively increase physical tolerances, stamina, endurance, and productivity with the eventual goal of improved employability.[7(p 139)]

As a treatment program, work hardening uses structured work simulation tasks to accomplish several goals, depending on the needs of the injured worker. Work hardening results are achieved through the use of specific vocationally oriented treatment strategies. A few of the most important are discussed below.

**Goaling Process**

Goals are an important part of the identification and treatment of symptom magnification syndrome. The term *goaling* is defined as the "process of identification and development of a primary goals hierarchy."[7] If you ask a symptom magnifier, "What are your goals right now?" you may get one (but not several) of the following responses:

- To maintain my temporary disability payments.
- To find/get a job.
- To get out of the house and away from my spouse, in-laws, and the kids.
- To get my spouse, in-laws, attorney, and others off my back.
- To find out what I can do.
- To find/get a cure for my problem.
- To get rid of my symptoms.
- To return to work.
- To earn what I was earning before I got hurt.
- To get my prescription filled.
- To be my own boss.
- To work outdoors.

Taken individually, each of these goals is inadequate and not sufficient to provide guidance and a future orientation. What is needed is a listing of several goal statements, each of which is a clearly developed and complete communication about that aspect of the person's life that the goal addresses. Goaling is important because it expands the symptom magnifier's sense of control over his or her life. Goaling assists the symptom magnifier to establish a future orientation, to develop a shared reality base, and to begin to receive positive feedback from his or her social network.

A goal is defined as a distinct, complete, and clear communication about one factor that makes life more satisfying. Goals fit within a context between purpose and objectives:

PURPOSE
\
GOALS
\
OBJECTIVES

Purpose is what gives life meaning. Many people are not consciously aware of their purpose. Some people who are consciously aware of their purpose have accepted purpose offered by others (eg, religious and political leaders). The development of purpose is a task that begins in early adulthood and for many is never completed. Purpose is often difficult to know. Purpose may be great or small, expansive or narrow. Purpose affects the direction of life through its effect on a person's selection of goals. A person can get closer to an understanding of his or her purpose by moving toward a full understanding of his or her goals. Goals can be thought of as helping to "form" and define purpose.

Objectives are the practical implementation of a person's goals. Objectives are statements about what one *can achieve* whereas goals are statements about what one has the *potential to achieve*. Objectives should be realistic, whereas goals, as one of my injured workers once told me, should have "realistic potential." If a person can think of himself or herself as having an "expectational horizon"—that is, a range within which the person expects to perform or achieve—objectives are set *at* that horizon whereas goals are set *just beyond* that horizon.

The goaling process is self-contained and has no performance criteria other than completion. As such, it is an excellent graded mastery task. Goaling is an exercise that functions to assist the symptom magnifier to develop the frame of reference of the worker with a future orientation. This technique does not identify a specific job, but gets at the symptom magnifier's values and desires. Goaling is actually a form of cognitive therapy that is intended to change the symptom magnifier's opinion of himself or herself.

**Samurai Game**

Another useful approach to the graded mastery treatment of symptom magnification syndrome is the samurai game. Here, the symptom magnifier

- pledges to keep every promise that he or she makes.
- maintains a diary in which each promise that is made throughout the day is recorded.
- reports to the therapist on a daily basis.

If the patient has not kept every promise during the preceding 24 hours, the patient must "commit" hari-kari. Of course, hari-kari is not *actually* required. The point of the exercise is to emphasize the importance of keeping promises.

In samurai, the issue of the person being responsible for control of his or her life is brought to the fore. That is, the ability to keep promises is based on a combination of the number of promises made, the difficulty of the promise, and the commitment made to keep them. If a person wishes to avoid hari-kari, making promises can be avoided entirely. If a person has a low level of commitment, a few simple promises can be made. As a person's level of commitment and self-belief improve, the promises can become more numerous and more difficult to keep.

**Work Function Development**

"Work function themes" are the rules (usually unstated and unconscious) that are used to guide participation in work activities. The work function themes for anyone can be known, verified, and modified to become more appropriate (or congruent) with the person's true capabilities. Clarification of the injured worker's tolerances improves employability by allowing better definition of the jobs the injured worker is able to perform.

One method for the identification of work function themes has been developed by the author using the *WEST Tool Sort*.[6] The *Tool Sort* is a commercially available test in which the symptom magnifier sorts through a deck of cards on which there are illustrations of tools in each tool's typical application, one tool per card. In this process, the symptom magnifier is instructed to "look over each card and place it in one of five piles," according to the symptom magnifier's perception of his or her ability to use the tool. After the symptom magnifier sorts the cards, the evaluator reviews the sort in one (or more) of several ways:

1. Sort is recorded as initially conducted by the symptom magnifier. This often gives an excellent indication of the symptom magnifier's magnification of his or her disability in terms of its impact on ability to work.
2. Sort is reviewed with the symptom magnifier and challenged by the evaluator based on the evaluator's understanding of the symptom magnifier's disability status. This is an excellent source of information concerning the symptom magnifier's current activity level and his or her understanding of the impact of the disability.
3. Sort is reviewed by the evaluator to educate the symptom magnifier concerning appropriate expectations in terms of tool use given the disability. (This has been formally developed by Patricia Smith as the "Perception Validation Interview."[8])
4. Sort is reviewed by the evaluator to match tool use abilities to job demands and to identify:
    - tools that require modification;
    - the way in which work hardening activities, if required, should be focused;
    - whether or not a return to these work activities is possible, and if so, how job modification will need to be conducted.

The use of hand-held tools is an important part of most workers' repertoire of skills. Illness or injury that has detrimentally affected the physical capacities that underlie these skills will have tremendous negative impact on workers' employability. More important, injured workers' reports of their use of tools provides valuable clues to their perception of their capacity.

**Symptom Control Development**

Through the use of work pacing, proper body posture and body mechanics, and the substitution of productivity for symptoms as a method of self-assessment, symptoms are controlled (not necessarily decreased) and made much more predictable. The injured worker is able to work around the symptoms as they become more predictable because they are recognized as being tied to certain job tasks that he or she must perform.

The injured worker learns the relationship between job tasks and symptoms and is able to work around or modify the method by which these job tasks are accomplished.

## Tool Adaptation or Job Modification

Tool adaptation and job modification are necessary components of every work hardening program. As a primary course of action in a rehabilitation program, these strategies have been shown to be both effective and inexpensive.[9] However, formal job modification is undertaken much less often (probably more because of rehabilitation professionals' attitudes than employers' resistance), and the author believes it can be beneficial.

Tool adaptation and job modification occur in almost every job, whether or not the worker is disabled. In rehabilitation, experience has shown that most modifications are developed by the injured worker working in his or her own work environment without benefit of the professional's input. Work hardening gives the professional an opportunity to work with the injured worker in a laboratory setting. The work hardening environment, because it uses work simulation tasks with the injured worker on a daily basis for several days in a row, allows experimentation with different job and tool modifications. Thus, by the time the injured worker goes out to the workplace, most of the bugs have been worked out.

## Reconstruction of Worker Role Identification

In treatment of symptom magnification syndrome, it is important to note that conditions (even in those conditions in which reinforcement occurs) will cause depression if these conditions are independent of the person's prior response. That is, people who are reinforced indiscriminately will not be assisted to pull out of their syndrome. The key for developing a reconstituted sense of self and recapturing competence is the perception by the person that his or her own actions control the experience. To the degree that uncontrollable events occur, ego strength will be undermined and the symptom magnification syndrome will continue. To the degree that controllable events occur, a sense of mastery will result. Graded productive behavior directly competes with symptom magnification syndrome behavior. Meaningful work can be used therapeutically to build a sense of being useful and being able to make a contribution. To reconstruct the worker role, the career development sequence is recapitulated. Tasks are selected that produce symptoms but that can be completed. The end product has reinforcement value (eg, bringing home a child's toy for the family). The demands for quality and concentration are graded incrementally and the symptom magnifier is reinforced for each new level of success in the worker role.

There are certain type-specific treatments available. The approach for the type 1 refugee involves practical conflict analysis and resolution. In this approach, the symptom magnifier is assisted to analyze and resolve the conflict situation that may require professional family counseling.

The approach for the type 2 gameplayer involves unequivocal identification of the symptom magnification syndrome as part of the treatment. This eliminates evasion as an option. This is an important goal because when the symptom magnifier gets away with the behavior it is reinforced. In addition, unequivocal identification as treatment is the best justification for the identification process. If the *only* reason for testing is to expose the gameplayer, maximum voluntary effort testing may not be justifiable.

The approach for the type 3 identified patient includes a strategy called "disability expansion," which focuses on and includes other life roles that can be reclaimed. Goaling is performed with regard to these life roles (eg, "What do you miss doing with your family?").

In addition, for the identified patient behavioral control strategies are necessary for symptom magnifier safety and liability control, as well as for treatment. Remember that this is the person who may do things to hurt himself or herself because he or she is so comfortable in the patient role, and threatened by leaving it, that he or she will behave unconsciously to maintain it. Such problems are usually manifest as poor compliance with medication, body mechanics, and rules of safety or self-management.

Symptom magnification syndrome can be effectively treated. Although a large-scale study has not yet been conducted, clinical reports from centers that are using the strategies presented here suggest that as many as 50% of those people identified as symptom magnifiers can be treated effectively and will return to work.

## SUMMARY

The symptom magnification syndrome is a new concept that has important consequences for all who are involved in the industrial rehabilitation process.

For the patient, the syndrome negates the possibility of the type of involvement that is needed to benefit fully from rehabilitation and is effective in preventing the development of healthy behavior, the adjustment to a chronic disability, and the effective and full utilization of all health care services. Its successful resolution will allow the high quality health care that is available to be utilized fully.

For the employer and the workers' compensation insurance carrier, because the syndrome results in inefficient and poorly effective utilization of health care services, it adds enormous costs to an already expensive responsibility. Successful resolution of the syndrome will cut the cost of service and allow others who can benefit to have access to additional services.

For the rehabilitation professional, the syndrome results in confusion and frustration. Exposure of the professional to a substantial number of such patients will bring about a very real negative impact on the professional's morale and creativity, thus draining some of this society's most important professional resources.

## REFERENCES

1. Matheson LN: *Work Capacity Evaluation*. Trabuco Canyon, CA, Rehabilitation Institute of Southern California, 1982.

2. Matheson LN: Symptom magnification syndrome. Paper presented at National Rehabilitation Association meeting, Anchorage, AK, 1984.

3. Matheson LN: Work hardening. Paper presented at National Rehabilitation Association meeting, Des Moines, IA, 1985.

4. Seligman MEP: Helplessness: *On Depression, Development and Death*. New York, WH Freeman Co, 1975.

5. American Psychiatric Association: *Diagnostic and Statistical Manual of Mental Disorders*, 3rd ed. Washington, DC: APA, 1980.

6. Matheson LN: West Tool Sort. Huntington Beach, CA: Work Evaluation Systems Technology, 1982.

7. Matheson LN, Niemeyer LO: *Work Capacity Evaluation: Interdisciplinary Approach to Industrial Rehabilitation*. Anaheim, CA: Employment and Rehabilitation Institute of California, 1986.

8. Smith P: Perception validation interview. *WESTWORK* 1984;2(1).

9. California Workers Compensation Institute: *Vocational Rehabilitation*. San Francisco, CWCI, 1986.

## SUGGESTED READING

Matheson LN: *Work Capacity Evaluation*. Rehabilitation Institute of Southern California, 1982.

Matheson LN, Ogden LD: *Work Tolerance Screening*. Rehabiliation Institute of Southern California, 1983.

Matheson LN: Ogden LD, Schultz, KS, Violette K, "Work Hardening". *American Journal of Occupational Therapy*. May, 1985.

Matheson LN: *Symptom magnification syndrome*. National Rehabilitation Association, Anchorage, Alaska, 1985.

Matheson LN: *Work hardening*. National Rehabilitation Association, Des Moines, Iowa, 1985.

Matheson LN and Niemeyer LO: *Work Capacity Evaluation: Interdisciplinary Approach to Industrial Rehabilitation*. Employment and Rehabilitation Institute of California, 1986.

Seligman MEP: *Helplessness: On Depression, Development and Death*. New York, WH Freeman Co, 1975.

# Section 19

# Chronic Pain Management

*Jodi Landis West*

## WORK INJURY AND CHRONIC PAIN

The workers' compensation system is responsible for providing a work-injured employee the opportunity for medical treatment to restore the highest level of functioning possible. Once the employee's progress has reached a plateau, a final settlement is often made to the injured worker. Regardless of the standards used in determining the final compensation settlement, ideally it should be based on loss of function, not on the employee's subjective manifestations.

Issues become clouded when the injured worker fails to show progress and exhibits significant limitations secondary to his or her complaints of pain. Should the compensation settlement then be based on the person's degree of functioning even though the reason for his or her deficits is based on subjective reports of pain?

When the injured worker has received extensive rehabilitation, yet continues to show poor to no improvement, the workers' compensation carrier may turn to a pain center for help in sorting out the magnified symptoms from the objective deficits. The typical worker enrolled in a pain management program has already spent an extended time in the medical system. Therefore, when he or she is enrolled in a pain management program, the problem is not diagnosis, but rather, the failure to respond to treatment.

The goals of an injured worker's participation in a comprehensive pain program are:

- To maximize the person's level of function.
- To teach the person to manage his or her pain.

- To reduce the person's dependency on the medical system and promote the person's independence.
- To facilitate case resolution.

## CHRONIC PAIN ATTRIBUTES

Once pain has persisted more than 6 months, it is defined as chronic. Although factors that influence a person's perception of pain vary among people, patients with chronic pain present common characteristics, regardless of the mode of injury.

- The person is disabled beyond the scope of the organic disorder. Decreased endurance, strength, and mobility become the result of progressive inactivity.
- Time spent resting also is time spent alone. This social isolation can contribute to a person's perception of life as boring and depressing.
- Overt pain behavior is customary. This behavior can present in various ways; however, it is most commonly displayed as an antalgic gait, guarded movement, and heavy sighing and groaning. Excessive verbal pain complaints are the norm in the chronic pain patient population. In essence, pain becomes the person's central focus, and the person's behavior becomes a reflection of this focus.
- Preoccupation with pain often paves the way for the person to assume the "patient role" both at home and in the community. A gradual dependency on others results from this role as other people begin to take on the injured person's responsibilities such as household tasks and parenting. As a result, the person generally is cared for by those around him or her. This type of response from others reinforces the person's inactivity and pain behavior and, consequently, perpetuates the person's chronic pain-dysfunction cycle. Returning to normal responsibilities and activities thus becomes more difficult for the person as the patient role is ingrained.

## HISTORY OF TREATMENT MODELS

A multidisciplinary approach has not always been used in pain management. Pain management philosophies and pain center programs have evolved over the years. Currently, the multidisciplinary approach is being emphasized.

When pain centers originated, they presented a dichotomy in their treatment philosophies. Surfacing after World War II, the treatment approach of the pain center was based on either the physiological aspect or the psychological aspect of the person's pain. This dualism significantly decreased the issues addressed in the treatment of a person's pain. By placing a patient in one category or the other, the treatment would tend to focus on the psychological issues and overlook the physiological issues, or ignore psychological issues to treat physiological problems only. As is being discovered, there is overlap in the two components, causing the philosophy of a unidisciplinary approach to be increasingly questioned. This dichotomy—treatment of mind versus treatment of body—has not been rejected universally; however, medical science continues to make significant contributions to the theory that pain is multidimensional and thus requires multiple attention for effective treatment.

A unidisciplinary approach is seldom successful in treatment of chronic pain because chronic pain is a multidimensional phenomenon. A person's neurological, physiological, motivational, social, and cognitive make-up all affect his or her perception of pain.

When chronic pain and its contributing factors are involved, a team approach by professionals from various disciplines is most effective in treating the person's pain. The team is generally headed by a supervising physician (often a physiatrist) or a psychologist. Included typically are a physical therapist, occupational therapist, social worker, nurse, biofeedback technician, recreational therapist, and a vocational counselor. Input from the diversified team members provides a holistic approach to pain management.

The overall goal of a pain management program is restoration of function to its maximal level. Each discipline involved in treatment must derive specific goals after a complete evaluation of the person. A problem list of conditions blocking the person's functional progress is then compiled. Specific goals of treatment are formulated and should be designed to provide the person direction in resolving problems and increasing the likelihood of successfully reaching maximal level of functioning.

Many times patients with chronic pain are unaware of "problem" areas that contribute to continued pain and dysfunction. The treating professional must provide adequate education and counseling so that the person can identify the problem as a contributing factor. Once acknowledgment occurs, short-term goals can be designed in a progressive fashion in an effort to diminish the contributing factor.

## PROGRAM PLANNING

Unfortunately, not all patients with chronic pain benefit from a pain management program. Studies have been done in an effort to determine which patients are most likely to benefit through participation in a pain management program. Results of these studies remain inconclusive. A greater likelihood of success has been noted when patients possess certain characteristics; that is, when the patient is willing

- to commit an extended period of time and effort;
- to participate in all aspects of the pain program; and
- to abstain from narcotics and alcohol.

The success rate is improved when the person also has an adequate family support system. The person who possesses sufficient insight to recognize factors contributing to his or her pain and has the desire to reverse reinforcing situations is the best candidate for success. Conversely, the person who is content in the patient role and receives adequate financial support to meet daily needs shows the lowest success rate in a pain management program; the desire for changing the situation is absent.

### Goals and Methods

Change is necessary if a person is expected to improve functionally. Changes in the areas of posture, movement, eating, sleeping, medication, and social interactions may need to take place for the person to progress. The responsibility for making these changes must be assumed by the person; the role of the treatment team remains one of providing information and guidance.

Once evaluations of the person have occurred and his or her appropriateness for the program has been determined, the person is informed of the team's expectations while in the program. A contract outlining responsibilities in the areas of attendance, medication, nutrition, activities, rest, and goal attainment can effectively provide program guidelines. It is vitally important that treatment goals be stated explicitly and that verbal reinforcement occur regarding the primary goal of restoration of maximal functional level. The client should understand that although the goal is decrease in pain, it may not happen. Rather, success is measured by gains made in functional capabilities, not by the person's reported pain level. The contract should be written in lay terms and presented to the person by all team members in a conference setting. Specific treatment goals also can be presented at this time by each discipline. Failure to adhere to contract guidelines, after adequate warning, should result in dismissal from the program.

The patient is not the only one with responsibilities. The team members have obligations to the patient as well. Objectivity in evaluation and treatment is the foremost responsibility of the health professional. Ongoing assessment and progression of the person's treatment also are important. Communication between the person and the health professional will facilitate progress. It allows for appropriate modification of goals and treatment in response to the person's needs. It also provides an opportunity for reinforcing positive feedback when goals are achieved and healthy behavior is demonstrated by the person.

Treatment team members also have responsibilities to co-members. Timely and accurate sharing of a person's progress is mandatory if the multidisciplinary approach is to be successful. Information should be given both formally in scheduled conferences and informally between disciplines as needed. It is important for behavioral or motivational problems to be brought to the team's attention so that appropriate action can be determined and reinforced.

## COMPONENTS OF A PAIN MANAGEMENT PROGRAM

### Education

The foundation of a pain management center's treatment is education. Education and the insight that follows increase the person's independence from the medical system. By learning the factors that perpetuate pain and dysfunction as well as ways to address these factors, the person can develop self-sufficiency in coping with pain and improving function.

As with all cases of learning, multiple, repetitious exposure to a topic is necessary. Group and individual settings are encouraged. However, lectures alone are insufficient in teaching the person; there must be supervised time in which to practice new techniques.

There are several areas in which the person must be educated. The physician's responsibility is to explain the diagnosis and its implications. Anatomy, proper body mechanics, good posture, work simplification, and pacing skills are topics covered by the physical therapists and occupational therapists. The nursing staff educates regarding medication, nutrition, and weight loss. Nonphysical factors contributing to the person's pain, as well as the consequences a person experiencing chronic pain may face, are addressed by the team psychologists and social workers. Sleep disturbances, pain coping techniques, biofeedback, stress management, relaxation techniques, pain behavior, and communication techniques are addressed by the team psychologist. The social worker not only serves as a liaison for the injured worker and his or her employer, but also provides information and counseling as it pertains to vocation rehabilitation, family readjustment, and the meaning of disability from a legal, medical, and individual standpoint.

**Physical Intervention**

Whereas education is the foundation of a good pain management program, functional restoration is the framework. Based on physical therapy and occupational therapy initial evaluations, deficit areas are targeted by an aggressive physical restoration program.

Decreased mobility, strength, and endurance often are present and relate to poor body mechanics and posture. Proper movement in both occupational and recreational activities and safe body mechanics must become second nature.

The physical therapist should design specifically for both the person and the group regarding exercise formats. Beginning at a submaximal level, exercise for the patient with chronic pain should be progressed along targeted dimensions. The person should be informed of daily and weekly exercise goals.

Low intensity, long duration stretching is employed to stretch painful tight muscles and regain flexibility. An increase in muscle flexibility may have the end result of decreased pain as well as improved motion. The person's stretching exercises should be designed to target tight muscle groups. Common areas of tightness include hamstrings, back extensors, upper trapezius, and pectoral muscles.

Another area of importance is the person's cardiovascular endurance. Because of decreased activity, decreased endurance is prevalent in the chronic pain population. Because fatigue increases pain, improved endurance may help decrease pain during activities. A training circuit using walking, stationary bikes, and a low impact aerobics program often is used.

A final area is strength. Depending on equipment available, the physical therapist may use isometric exercises, isokinetic exercises, or isotonic exercises. Regardless of strengthening techniques used, beginning at submaximal levels and gradually increasing difficulty along defined goals is the optimal method of progression. An option is inclusion of stretching, strengthening, and endurance activities as part of a water exercise program.

The relaxation properties of water in conjunction with its warmth often allow improved stretching at a greater comfort to the person.

Many people enrolled in a pain management program find the outlined strengthening, stretching, and endurance programs strenuous. An increase in both pain behavior and pain complaints often are the result of beginning a physical restoration program and may serve as a roadblock in a person's progress.

One behavior modification approach used is ignoring pain behavior in an effort to extinguish it and providing positive feedback for behavior or exercise that is normal and healthy. Token rewards in conjunction with liberal verbal praise can be given on the person's achievement of a goal. Rest periods can be scheduled either on a time basis or on an activity-completion basis, but not on a pain contingent basis.

As a person's endurance, strength, mobility, and psychological tolerance improve, a work hardening program may be introduced. (See Section 15, *supra.*)

Occupational therapists encourage the person to participate in functional activities. Lectures and group discussions covering utilization of leisure time, alternative forms of recreation, and readjustment to the community and peers are valuable. Periodic excursions are planned. These outings may be to an athletic event, a movie, a museum, or a local attraction. Through these outings and group discussions, the person is reintroduced to socializing and pursuing recreational interests.

### Psychological Intervention

Physical restoration, in itself, has a small chance of success without some form of psychological intervention. This can aid the person in addressing issues that contribute to the pain process. The psychologist can also offer assistance in dealing with the consequences of the person's injury.

Most pain management centers incorporate individual and group psychotherapy. Customarily, individual relaxation training and biofeedback are provided. Primary goals of biofeedback and relaxation techniques include decreased pain, relaxation, and stress management. Initial biofeedback sessions are focused on educating the person on the relationship among stress, tension, and pain. Once this cause and effect cycle is established, the person is taught ways to decrease muscle tension and thereby reduce pain.

By learning to manage pain through biofeedback and other non-drug techniques (eg, ice, heat, relaxation, imagery, or transcutaneous electronic nerve stimulator [TENS]), the person increases his or her independence from the medical system and improves the chances of achieving the overall treatment goal of restoration of function.

### Coordination of Services

Another team member working closely with the injured worker is the social worker. Social service often coordinates conferences, functions as a patient advocate, and helps maintain good communication between the injured worker and the workers' compensation carrier. The social worker may provide counseling to the person, especially as it pertains to vocational retraining and re-entry into the work force.

On completion of a pain program, progress of the person and the outcome of the treatment program are documented. Each discipline reevaluates the person and compares final data to base line measurements. Functional capacity testing is invaluable at this juncture. It provides the person, the physician, and the employer with the person's maximum weight capacities and job restrictions based on limitations determined through this objective testing. Case resolution is facilitated either by return to work or by determination of residual objective deficits.

The person should receive a home program for continued strengthening, flexibility, and aerobic conditioning. Newly learned body mechanics, stress management techniques, and pain coping techniques also are taken with the person. Application of this knowledge outside the confines of the pain center can enhance an active and productive life style. This in the end is the true measure of success of a pain management program.

## PAIN CENTER EXPERIENCE

### Patient Data

A study of the 8-week outpatient pain program at Indiana Center for Rehabilitation Medicine (ICRM) was done in 1986. Results of that study include:

- A positive outcome was obtained by 64% of the patients treated ("positive outcome" defined as returning to work for the same employer, either to the same job or to light duty).
- One year or less after injury, 79% success rate.
- For 13 to 24 months after injury, 39% success rate.
- For more than 2 years after injury, 33% success rate.

Age and marital status were not statistically significant variables. These statistics include patients who started the program but dropped out before completion of the full program.

Typical diagnoses for patients enrolled in this program were lumbosacral sprain, myofascial pain syndrome, failed disk syndrome, and reflex sympathetic dystrophy.

**Program Specifics**

The protocols for pain center management at ICRM have been condensed into a 6-week program. The following is a typical protocol.

On referral of an injured worker by a workers' compensation carrier, the person undergoes a comprehensive, multidisciplinary assessment over the course of two days to determine appropriateness for the program and to aid in treatment planning. Assessment includes medical tests, such as an electromyogram, and x-ray films, if current tests are not available. A physical therapist assesses the person's strength, flexibility, endurance, posture, gait, and trigger points. A personality evaluation is performed, which includes an interview by the psychologist and administration of the *Minnesota Multiphasic Personality Inventory (MMPI)*, the *Sickness Impact Profile*, the *Back Depression Inventory*, and the *Spielberger State Anxiety Inventory*. The social worker's evaluation of social supports, including family relationships, the Work Environment Scale, and a Sabotage Scale also are performed. Biofeedback base lines to determine excess muscle tension are performed initially. The nursing staff obtains a complete medical history of each person, which includes all medication the person is currently taking. Once all this information is gathered, the treatment team then meets for an initial conference to which the workers' compensation representative is invited to attend. Appropriateness for treatment is discussed and recommendations are made. A problem list is developed and used for goal setting. Goal setting occurs with the patient and team members present.

During the first 2 weeks of the program, the person attends classes concerning posture, the basics of workers' compensation, sleep disorders, body mechanics, and the physiology of pain. He or she receives individual physical therapy (primarily modalities and instruction in strength, flexibility, and endurance exercises). The person has attended group discussions covering nutrition and establishment of a diet; biofeedback training and group relaxation; recreational therapy excursions; and swim exercise program. Also, the person has the opportunity to have individual consultations with the physiatrist, nurse, psychologist, and social worker. Specific and realistic goals are determined at this time and a timeframe for achievement is established.

By the third week, the person is advanced from physical therapy modalities to therapeutic exercises and work hardening activities. Body mechanics and work simplification techniques are stressed. Individual biofeedback and counseling with the psychologist as well as with the social worker continues; but most exercises, education, and consultation are now done in a group setting. The day begins with a check-in with the nurse who reviews each person's eating log, discusses any medical issues with them, and conducts a weigh-in on Mondays and Fridays. The day ends with a group relaxation exercise led by a psychologist. Recreational therapy continues throughout the remainder of the program.

Increasing amounts of time and effort are spent by the person performing work hardening activities and endurance activities. Continued individual time is spent with the team psychologist and social worker. Daily check-in and group relaxation exercises continue. By the end of the 6-week program, each person is in work hardening or other therapeutic exercise approximately five to six hours each day.

Throughout the program, the staff is liberal with encouragement and praise. In each area, stars and other tokens are regularly earned for progress toward a goal. A chart is posted in the patient lounge on which medication use and weight loss or gain is displayed. The patients are encouraged to use non-drug techniques for pain control.

Peer support has been found to be of benefit in treatment. A "Patient of the Month" is chosen by the treating team. A framed certificate is awarded and a banner is displayed in the exercise area to commemorate the event. The emphasis is on positive gains and attitudes, with pain behavior and other negative activities discouraged.

At the midpoint and at the end of the program, the person's total functioning is reassessed by all disciplines. A formal case conference is held, which the patient and the workers' compensation representative are invited to attend. A functional capacity assessment (see Subsection 3 of Section 14, *supra*) is administered at the end of week 6 of the program. Case closure is based on input from the treating team, the employer, the third party payer, and the patient. The patients completing ICRM's pain management program are (1) returned to work, (2) given a permanent partial impairment rating, or (3) referred to vocational rehabilitation.

**Case Study**

A 33-year-old female licensed practical nurse injured her lower back when transferring a patient from a stretcher to a bed. She complained of midline lower back pain, left buttock pain, and pain radiating down her left posterior thigh. She also reported left lower extremity weakness. Pain was reportedly increased with all activity. She reported "rest" and heat as decreasing her pain. Sleeping difficulties also were noted.

Diagnostic testing revealed no sign of organic disorder. Physical examination revealed patient as being obese with poor posture. Palpation indicated muscle tightness bilaterally in lumbar paraspinal muscles. Trunk mobility was limited. She presented normal leg strength bilaterally, but demonstrated decreased stance on left lower extremity, decreased hip excursion, and decreased heel strike on left during gait.

During her psychological evaluation, the patient disclosed that she did not understand what was occurring physically and was sure "there's something in there that isn't right." She discussed her pain at great lengths and gave graphic details as to how it felt. She continued that she spent much of her time sitting around thinking about what she could not do. She stated her self-confidence and self-esteem had deteriorated since her injury. Moderate to excessive pain behavior was noted in all disciplines' evaluations.

The patient was determined to be an appropriate candidate of the pain management program. Specific goals and a performance contract were presented to the patient.

The patient successfully completed the 6-week program; however, her progress began slowly. She initially exhibited an increase in pain behavior and pain complaints during her first 4 weeks. Pain behavior was ignored by the treatment team. Increase in activities and attainment of daily goals, no matter how small, were given attention and praise. Intense psychological counseling continued, and as the patient began showing signs of insight as to the secondary gains associated with her pain behavior, she began to demonstrate less pain behavior. The patient demonstrated minimal to no pain behavior during her final week of the program. Her functional capabilities and limitations were determined by performance of a functional capacity assessment. The patient's physical capabilities were insufficient for her to return to her previous job. However, she was able to continue to practice as a licensed practical nurse in an outpatient department of the hospital in which she was employed.

The case study exemplifies a person who has no sign of organic deficit yet continues to experience pain. Although the patient was unable to return to her previous job, she was able to return to the work force in her chosen profession. A physical restoration program in itself would have failed without comprehensive psychological intervention. Likewise, without a competent physical reconditioning program the patient's chances of returning to the work force would have been slim.

By addressing both the psychological and physical components that contribute to the pain process, the injured worker with chronic pain is being provided a way to help himself or herself attain a more self-sufficient and productive life style.

**SUGGESTED READINGS**

Aronoff GM, Evans WO: Evaluation and treatment of chronic pain at the Boston Pain Center. *J Clin Psychiatry* 1982; 43:4–9.

Grabois M: Pain clinics: Role in the rehabilitation of patients with chronic pain. *Ann Acad Med* 1983; 12:34–39.

Indiana Center for Rehabilitation Medicine: *The Occupational Injury Rehabilitation Program* (brochure). Indianapolis, IN: Indiana Center for Rehabilitation Medicine, 1987.

Schofferman J: Management of chronic pain, in White AH (ed): *Failed Back Surgery Syndrome*. Philadelphia, Hanley and Belfus, Inc, 1986, pp 115–127.

Unikel PI: How we learn chronic pain and sickness, in Brena SF (ed): *Chronic Pain*. New York, Atheneum Press, 1978, pp 19–25.

Unikel PI, Chapman SI: The pain prone patient, in Brena SF (ed): *Chronic Pain*. New York, Atheneum Press, 1978, pp 27–33.

# Part IX

# The Return-to-Work Process

Each professional in the return-to-work process may consider himself or herself as the main link in this dynamic process. However, soon the professional realizes that there are many people involved. They all have their own point of view, their own expertise, and a different relationship with the injured worker. The process actually is centered around the injured worker and how the work injury affects progress.

Understanding all the points of view in the system strengthens each person in his or her professional role. In the sections in Part IX, experienced professionals present clarifying perspectives of their involvement in the progression toward work return for the employee who is injured.

## Section 20  Return to Work: The Physician's Role

The role of the physician has changed dramatically over the years. Formerly, medical and legal requirements put the physician at the helm of the medical case. This was a very responsible role for a professional whose main concern was the promotion of healing of the worker. It also was a role that sometimes lacked objectivity because of the physician's need to be a patient advocate.

Currently, the physician uses medical skills both in healing and in evaluation. Thorough reports indicate to the other professionals involved the extent and outcome of the medical injury. The physician no longer needs to bear the total responsibility for return to work. Interaction with the other professionals allows a stronger physician role in the medical side, but a less demanding role in making vocational judgments.

## Section 21  The Rehabilitation Consultant

The use of the rehabilitation consultant in prolonged work injury cases is increasing. This neutral professional facilitates functional work return at the safest possible level.

Both vocational professionals and medical professionals (often nurses) act in the role of rehabilitation consultant. They interact with the employer and medical specialists attending the worker to determine the most advantageous case management for all concerned.

## Section 22  Employer Point of View

Although the employer always needed to take charge of the injured worker on the return to work, he or she was often the least informed. When the return to work was accomplished it was assumed that medical problems were no longer present. Therefore, for years the employer put people back to work without a full understanding of medical constraints or prevention of reinjury.

In the current model, the employer can be pro-active in setting the stage for positive employee-employer relationships. When this occurs, there is already a positive attitude when an injury occurs, and the employee and the medical professionals do not see the need to "take sides."

A safe and timely return to work assures better productivity for the employer and allows the employee to be more protected from reinjury.

## Section 23  The Therapist

Physical therapists have always treated work injury cases in a clinical manner. There is now a stronger role in dynamic physical assessment and restoration of the worker. Occupational therapists use their skills in functional activities and work simulation to add to the functional restoration process.

With the goal of return to work, therapists have developed new skills and techniques to bring the injured worker past the point of physical healing and into functional usefulness.

## Section 24  The Attorney's Role

There are two sides to every question; and when legal problems arise, there are two types of attorneys for each case. They may have completely different points of view and each feel they are serving the work process in the best manner. Viewing the opposing viewpoints shows a total picture of the workers' compensation system.

## Section 25   The Main Feature: The Employee

The employee is the most important person in the return-to-work process. The pressures and directions of all professionals represented in Sections 20 to 24 greatly affect the injured worker. The employee is the center of the action, and yet often is the one with the least amount of power. The employee must rely on others to determine his or her progress, as well as the status of his or her situation.

The more current enlightened model of work injury management encourages the employee to be in a central role, taking responsibility for physical healing and becoming pro-active in the return-to-work process.

# Section 20

# Return to Work: The Physician's Role

*William P. Fleeson*

## THE OCCUPATIONAL PHYSICIAN AND THE COMPANY: PARTNERS

One of the most important things an employer can do is to have a physician evaluate an injured employee as soon as possible. The first day of an injury is critical, because the sooner good care is started the better the chance the period of time before the employee can again work will be minimized.

In most cases the thoroughness and correctness of the initial evaluation will determine the entire subsequent course and progress of the case. It is when the injury is fresh, when the details of the circumstances are clear in the patient's mind and treatment with over-the-counter drugs or "home treatment" has not yet begun, that the physician can make the most accurate assessment of the injury. Early evaluation also means the patient can be included as a partner in his or her own care plan.

One reason for using an occupational medicine specialist is the treating physician needs to be familiar with the patient's workplace. The physician with this knowledge will be able to understand more accurately the patient's duty status and to estimate more correctly when the patient will be able to return to work.

### How To Find an Occupational Physician

Occupational physicians are medical specialists; that is, physicians who have specialized in either occupational medicine or preventive medicine, who have special training in the medical problems of the workplace, and who almost always have training in administration and public health with degrees in public health as well as medicine. Physicians with this specialty are generally skilled in evaluating almost any type of work injury and must have particular expertise in musculoskeletal problems because of the frequency of these injuries in the workplace. (Occupational medicine specialists can be found by contacting the American Occupational Medicine Association, Arlington Heights, Ill.)

An occupational physician may have any of several types of relationships with an employer. In large companies he or she may actually be an employee. The small company may contract his or her services as its consulting medical director, medical director, or company physician. In many cases occupational physicians have a private medical practice and treat patients from a large number of employers in their region.

An occupational physician, by virtue of his or her knowledge of the workplace, has a relationship with the employer that allows frequent communication regarding the patient's diagnosis, status of the injury, and expected return to work. The occupational physician's responsibility is to provide the employer with an objective evaluation of the injured worker's condition and to give the injured worker the best treatment possible to facilitate his or her return to a normal life as soon as possible. The occupational physician is very cognizant of the employer's facilities, policies, needs, and restraints.

In some cases, an employee may understandably wish to see his or her own physician, rather than one specified by the company. For the most part this also can lead to rapid return to work.

In either case, whether the family physician or the occupational physician evaluates the patient, early and close attention to recovery and return to work will hasten healing time and return to work.

## CORRECT DIAGNOSIS: THE FOUNDATION OF CORRECT CARE

Once an injury has occurred, the fundamental basis of every subsequent step is the establishment of an accurate diagnosis. Just as a slight mistake at the beginning of a rocket flight to the moon can result in a 10,000 mile miss, an inaccurate diagnosis at the beginning of treatment can lead both patient and physician down the path of incorrect care, poor progress, retesting, and delay.

The physician can usually form an initial clinical "impression", which is then the basis for initial and subsequent diagnostic work to pinpoint the diagnosis in a fairly short time. The diagnosis then forms the basis for appropriate treatment and for estimating the time that healing will take and the time likely to be lost from work. In addition, if the diagnosis is not established quickly, the physician will be unable to inform the patient about what is wrong with him or her and to explain it so that the patient's help can be elicited in treatment and rehabilitation. Incorrect diagnosis will mislead other members of the medical team, including physical therapists and rehabilitation professionals. Without correct diagnosis the employer also will be misled as to when the patient will return to work. Incorrect diagnosis will lead to more time lost later while further diagnostic measures are taken.

Finally, correct diagnosis at an early stage of the injury or illness will avoid further unnecessary testing. Such testing is expensive, sometimes unpleasant, takes a great deal of time, and occasionally carries its own risks.

The classic example of an incorrect diagnosis is the construction laborer who lifts a heavy object and has back pain. He might see a practitioner who diagnoses "muscle strain." Even worse he could receive the meaningless "diagnosis" that his "vertebrae are out of place." With these well-meaning but inaccurate assessments of the problem, the patient might receive treatment that is at best ineffective and at worst harmful for many weeks or even months. He may eventually find a physician who correctly diagnoses a herniated disk, nerve root impingement, or a spondylolysis. But by the time the correct diagnosis is made, the patient will have had several months of aggravating treatment, physiological deconditioning, and chronic muscle spasm because of pain and will respond slowly to appropriate treatment when it does finally begin.

**Diagnostic Workup**

The diagnostic workup for an on-the-job injury differs in some respects from a standard medical workup. The physician obtains a detailed history of the injury, with particular attention given to the position the patient was in at the time of injury, the weight he or she lifted or the object he or she was struck by, and the symptoms at the time of injury. A physical examination can be general but also emphasize the injured area or body parts, such as right shoulder, left knee, low back. These steps will generally give the physician a good idea of the diagnosis and the prognosis.

Further diagnostic work may include simple observation, especially as the patient leaves the office and goes for further testing. X-ray films are often useful, certainly in any traumatic injury. (The controversy over the value of low back x-ray films is too extensive to include here.) Laboratory work is not useful in many on-the-job injuries, particularly musculoskeletal injuries; but other occupational problems such as exposure to toxins, causative agents of skin diseases, and respiratory irritants and inhalants, will require much more extensive laboratory work.

During the initial examination and subsequently, manual muscle testing for strength and range of motion is obviously very important. Neurological examination and gait and strength testing on various parameters also is frequently of great use. With some workplace injuries, mostly nonmusculoskeletal injuries, other testing may be done. For example, an employee with a skin eruption or irritation of undetermined cause might require testing for various allergic phenomena that would be primarily based on laboratory work. In addition, the physician may be required to visit the worksite to understand various procedures and exposures.

As a case example, consider a patient who worked in a nursing home and had an undiagnosed skin rash on his hands and forearms for more than 2 years. He always improved when he was away from the worksite for 2 or 3 weeks, but then worsened within a few days of returning. No offending agent could be pinpointed. A physician visited the worksite and determined 17 possible chemical exposures; analysis of the ingredients revealed some had similar characteristics. Skin testing with five or six of the compounds then documented the specific chemical to which the patient was sensitized. Now, avoiding lotions and soaps that contain those compounds has eliminated his problem. In this case the problem may never have been solved had the workplace and its contents not been directly observed by the occupational physician, and subsequent evaluation been based on the observations.

Once the diagnostic work has been completed and it clearly pinpoints the diagnosis (ie, as much as is feasible at the initial visit), the physician should then formulate a treatment plan. Part of the treatment plan should be to determine treatment goals and to estimate the length of time that will be required for healing.

## BEGINNING OF THE TREATMENT PLAN

### Medical Care

Beginning treatment of a work injury as soon as possible is also critical. It goes without saying that the physician should choose treatment that is appropriate, conventional, and accepted. Treatment should be documented in the medical literature to be not only effective but safe, and it should have the clear purposes of relieving discomfort, allowing healing to take place, and restoring normal function and/or mobility of the injured part. There is no place in work injury management for unscientific "treatments."

A good physician always remembers the first rule, "Do no harm." For example, in many musculoskeletal injuries the initial treatment can be conservative and directed toward relief of pain and muscle spasm for a short period of time. During this time the body's own mechanisms will come into play and begin the healing process. This section is not the place to discuss specific treatments, medications, or therapeutic modalities. However, the treating physician obviously must be aware of the value of rest versus exercise, heat versus cold, passive range of motion versus active range of motion, splinting versus mobility, flexion versus extension, manipulation versus mobilization, muscle relaxants, therapeutic modalities, and the other standard components of a treatment regimen for this type of injury. For all injuries, the physician must of course be well versed in appropriate treatment, medication, and follow-up. When medical therapeutics are required (as they may be in many musculoskeletal injuries and back strains) the various nonsteroidal anti-inflammatory drugs, muscle relaxants, and analgesics should be used as indicated. These can and should be used in conjunction with restrictions on the patient's activities, and he or she should also be given instructions to perform or not to perform various exercises.

### Ordering Physical Therapy Modalities

In the treatment of common musculoskeletal and back injuries, application of standard physical therapy modalities is frequently warranted. The physician should have a thorough understanding of the various modalities, including their use, effects, and benefits. Once the physician has ordered therapeutic modalities, he or she should communicate frequently with the physical therapists.

As soon as the medical and therapeutic treatments have been initiated, the stage is set for adding "rehabilitation" components into the treatment program. First, however, the physician has an obligation to communicate with the employer.

## TRANSMITTING INFORMATION TO THE EMPLOYER

The employer wants the injured workers to be evaluated by a specialist who can do more than merely make the diagnosis and institute treatment. Also important is the determination of an injured employee's fitness for work. In making this determination, the occupational physician fulfills the role of advising the employer and helping the employer to plan. By communicating with the employer after the initial evaluation, the physician discusses the patient's prognosis, both short term and long term. In other words, the physician "translates" the medical diagnosis and treatment plan into parameters that mean the most to the employer: how long before the patient will be able to return to work and how much he or she will be able to do.

### The Occupational Physician's Responsibility

In most states, the workers' compensation regulations make transmittal of information about the diagnosis and medical status mandatory under law. Employers have a right to ask for the information, and physicians are obliged to provide it.

In addition, the physician has another related responsibility, and that is to provide an evaluation based on objective data. When advising the employer of an injured employee's medical status, suitability for work, and prognosis, those estimates must be based on the physician's objective findings tempered by his or her experience with the type of injury being evaluated. Knowledge of the workplace and the physical and psychological demands that will be made on the patient when he or she returns to work in whatever capacity should be considered. In making these decisions and determinations there is no "cookbook formula" that can be used; training and experience combined with knowledge of the workplace are assets.

Each employer is different, which complicates matters. Some employers simply want the physician to let them know when the patient's injury has healed. Other employers demand total control of the process, including exercising their right to approve all expenditures and referrals. Some employers will allow the injured worker to return for "light duty" during the healing period; others will adhere to the policy of "no return to work until he's 100%." Some employers attempt to second-guess the physician; others stay out of medical care completely and confine themselves merely to administrative concerns.

## Acute versus Chronic Cases

Early in an injury case, the physician must usually make the determination of whether the injury is acute or will likely be chronic. An acute injury would generally be defined as an injury that is likely to have a relatively short treatment period, and that will probably be resolved more or less completely if appropriate treatment is given. Examples of acute injuries include clean lacerations, minor fractures of the arm, a caustic burn on the skin, and a forearm muscle strain.

That a case will be chronic can frequently be determined immediately at the time of an injury. Chronic implies long lasting, with prolonged healing time and, possibly, a permanent disability as residual. An extreme example of chronic injury would be a fall from a power pole with transection of the spinal cord; it might have been only 2 minutes since this injury occurred but it can be immediately discerned that it is going to be a chronic injury. Acute minor injuries can become chronic, such as when what appears at first to be a trapezius strain results in long-term myofascial syndrome, with shoulder-hand syndrome, requiring extensive time off work and rehabilitation and leading to a persistent limitation of motion and pain. Whichever the injury is, the physician will be best advised to be aware of the distinction between acute and chronic early on and so advise the employer.

In many, if not most, cases it is most useful to share the same information and expectations about length of healing time with the injured employee. An intelligent, informed patient is a better patient, and he or she usually participates in the healing process much more actively when he or she has been given correct information about the injury. It also is most helpful when the patient understands what the physician expects in terms of the patient's role in the therapeutic program. As in communicating with the employer, it is again most important to communicate with the patient in terms of strictly objective findings. Otherwise, the physician can become mired in a pointless argument trying to determine how much pain the employee has, its effect on his or her ability to perform activities of daily living or at work, and so forth. Objective findings are just that. They form the most logical basis for decision making both on the part of the patient and on the part of the physician.

## ONGOING CARE

The main principles of ongoing care are (1) to use early on as many of the physician's diagnostic and treatment resources as possible; (2) to continue to observe the patient closely throughout the treatment program; and (3) to begin "rehabilitation" as soon as possible during the treatment program.

## Care for Acute Injuries

When a medical problem is most likely to be short lived and to respond well to a relatively brief period of treatment, the physician should see the patient as often as medically necessary to feel that he or she is on top of the case and doing the best by the patient. The patient should be seen with a great enough frequency that it can be easily determined when the care period is ending and when the time for early return to work on restricted duty is approaching. Most employers request that a final medical evaluation be performed when the patient is ready to return to work at full duty and that the physician so state in writing at that time.

## Chronic Treatment

Many of the same principles apply for a chronic treatment program. Re-evaluation intervals may be slightly longer and the physician may be able to rely on the reports from the physical therapist or other rehabilitation professional for progress reports. When the case is becoming extended, however, the physician will find it most valuable to keep in frequent touch with both the patient and the employer, so that both can be appraised of progress. This also will be valuable in detecting any danger signs, either physiological or psychological, that might be developing, during a prolonged recuperative phase.

## Rehabilitation at the Same Time As Treatment

In general, the best treatment outcomes occur when the dividing line between the period of treatment and the period of rehabilitation is blurred. By this the author means that treatment of most musculoskeletal injuries, and many other work injuries as well, should be blended imperceptibly into rehabilitation. In some cases, what would ordinarily be called rehabilitation should be started simultaneously with acute treatment. This means that rehabilitation should be viewed as a logical and natural step between treatment and recovery.

With all but the most life-threatening injuries, rehabilitation can start sometime in the first few days. With less severe injuries, such as low back disk injuries, physical rehabilitation should begin as soon as the patient is out of bed, most commonly within 1 to 2 weeks at the latest. With typical extremity fractures, severe sprains, or muscle tears, rehabilitation of the injured part should begin as early as consistent with safety; for the uninvolved parts, rehabilitation can usually begin immediately.

As an example, if a patient has a fracture of the humerus in the dominant arm, nothing will be gained by having that patient casted and home for the next 6 weeks, allowing the other extremities to atrophy. Active range of motion of the uninvolved extremities, simultaneous with general exercises such as walking, would be a typical example of combining early rehabilitation with early treatment. Experience has convinced this author that physical activity and motion and avoiding prolonged inactivity or bed rest are important in the treatment of most workplace injuries, particularly musculoskeletal injuries. Physical activity keeps injured employees toned up both physically and mentally. It also serves the purpose of emphasizing to patients that they still can be active. Such an exercise or rehabilitation program keeps the mind of injured workers focused on healing, and keeps the goal in front: they are working on their recovery so that they may return to a normal life and to work again soon.

In contrast, injured employees who spend their time being inactive begin to feel that they are "disabled," begin to focus on their discomforts more than on their health, and begin to feel that they in fact are an invalid. When this develops, such injuries take considerably greater time to heal, and generally the final result is not as satisfactory. Being physically inactive during the healing process of a musculoskeletal injury brings grave consequences, mitigating against an early and successful rehabilitation and return to work.

Because of this, it is most practical to start the plan for rehabilitation and recovery as early as possible. This plan should include treatments, rehabilitation modalities, and a graded return to work or light duty program. The physician can accomplish such a program by using a strengthening and/or fitness exercise program simultaneously. He or she can do this by enlisting the help of the physical therapists whenever available. Counselors and rehabilitation consultants, as well as facilities such as the YMCAs, fitness centers, and even motel swimming pools, can be used with success.

As an example, a simultaneous medical and fitness treatment program could be applied in a case of severe low back pain. After an appropriate brief treatment of muscle relaxants and physical therapy, during which time the patient had difficulty tolerating sitting and therefore was either lying down or walking, swimming could be added as early as the second week. Swimming is valuable because the patient is weightless and can provide his or her body with the appropriate mobility and aerobic exercise for the injury.

Next, as soon as tolerated, appropriate back exercises would be added. Then a light structured exercise program with equipment could be started. Not only does this make patients feel that they are an active participant in their treatment, but also they see on almost a daily basis that they are making progress and much of the fear of re-injury not only is resolved but is actually prevented.

An integral and critical part of such a program would be to include an educational component to give the patient intellectual skills about his or her back and how it works, correct body mechanics, and, most appropriately, performance of the job if it includes lifting, stretching, and reaching. Another immensely valuable aspect of a program is psychological support.

Beginning a program of physical conditioning as early as possible, although such a program might in other circumstances be considered strictly rehabilitation, helps the patient to constantly remember that he or she is engaged in a healing process that has the purpose of returning him or her to normal life. No longer is the patient a passive participant; he or she is an active participant with some self-direction and control over the outcome.

## ENDING CARE

The goal of rehabilitation should always be to have the patient reach the point when care can safely be ended in as short a time as possible. Any physician who unnecessarily allows a period of disability to be prolonged could only do so because he or she does not realize the consequences. The results of prolonging the period of disability are so detrimental to the patient, the employer, the physician, and virtually everyone involved in a work injury case that such an outcome must be avoided at all costs.

Keeping an injured employee off work for months, sitting around inactive, long past a point when return to work would have been possible, does the patient a great disservice that could have been avoided. First, early physical activity and return to work has beneficial effects, both physical and mental. Second, a person loses 10% to 20% of his or her muscle tone with each week of bed rest, and it does not take long before a patient is physically deconditioned. So, if a physician waits until the patient is completely free of pain, it may take so long that the patient will be inactive, soft and so prone to re-injury from minor stress and strain that the pain never will be gone.

A related factor to consider is that when a person is kept off work too long, it is almost inevitable that he will begin to focus on the little discomforts and aches and pains of normal life and will develop a fear of these because the physician has been "waiting until I'm pain free." Such a patient begins to think he or she is always being reinjured, which increases the fear of reinjury on the job. The result is that the point at which treatment can be ended will never be reached.

Therefore, the occupational physician may in some cases have to push a bit to overcome a patient's resistance to return to work, even light work, even though it is a logical part of the rehabilitation process. Some of this reluctance may be understandable. Unfortunately some also is simply a disinclination to get back into the grind of getting up each day and going to work. The longer a patient is off work, particularly with the facilitation of a well-meaning caregiver who does not resist keeping him or her off work, the more difficult it is to overcome this reluctance.

### Maximum Medical Improvement

Many employers or insurers will ask the physician when maximum medical improvement, or MMI, has been reached. In many states this is known by other names, but it generally means a plateau period where, to a reasonable medical certainty, no further medical treatment will effect a significant change in the patient's status.

Any physician should be cautioned that simply because all medical tests are normal does not mean that the patient is free of pain. The patient also should be educated that minor discomfort may be the last symptom to disappear and should be reassured that all professionals involved understand that he or she does indeed have pain. However, return of all of the testing to normal does indicate that the point at which the patient should be able to return to some kind of safe work, on a graded schedule, is close.

## SPECIAL PROBLEMS ASSOCIATED WITH WORKERS' COMPENSATION

Work-related injuries, and, thus, injuries that involve the workers' compensation system might superficially appear to be medically similar to non-work-related injuries, but they are not. They take longer to achieve a good result, they seem to be more frequently complicated by pain without objective findings, they are immensely more complicated by litigation, and they are almost always fraught with problems of secondary gain.

Studies consistently show that injuries associated with workers' compensation take two to three times longer to heal than the same injuries in a non-workers' compensation population. Many factors are responsible for this problem. There is a financial factor, for in many cases workers' compensation payments are tax-free and are in fact more money than what the patient was taking home when he or she was working. This creates the undesirable situation where an employee is rewarded for staying off work. There also are psychological factors, for when a breadwinner is out of work or disabled or in chronic pain, the effect is unpredictable but leads to prolonged recovery in many cases. Sometimes because the patient doesn't like or doesn't understand the medical system, his or her frustration and hostility also prolong the recovery. Hysteria and conversion reactions also are common in workers' compensation cases.

Sometimes it is the medical facts that unavoidably prolong recovery. For example, soft tissue injuries are not visible, take a long time to heal, and also are very painful. This is usually difficult for a patient to understand. This is made worse when physical deconditioning complicates the prolonged healing and leads to more pain. There are administrative factors in workers' compensation that build resentment against the physician, the employer, the insurer, and, sometimes, every professional involved; this also tends to prolong disability.

Frequently, healing is slowed by factors not at all related to the medical problem, such as anger against the employer and even a desire for revenge against the company. There also are psychological barriers to recovery. (See Section 17, *supra*.)

With respect to one specific problem in workers' compensation, it should be recognized that employers and unions could do their mutual employees and members a service by promoting the concept that workers' compensation payments are not "benefits": that money is "high pay for performing the difficult job of healing quickly."

## EARLY RETURN TO WORK AND LIGHT DUTY

One of the most valuable tools for return to work is a duty status that is variably called modified, restricted, special, adjusted, or light duty.

If the employer will adapt the workplace and administrative structures to allow restricted duty, the occupational physician has an invaluable opportunity to speed the patient's rehabilitation. Light duty does have many drawbacks but has generally been found to lead to early recovery and more rapid return to normal life and work.

An example of light duty that has been used for patients with forearm strains or tendinitis includes having the patient work in the parts room, even if he does nothing more than sort bolts while the strain is healing. This has led to an amazingly rapid recovery, for many reasons: the patient continues to be active, he continues to be an active part of the work force, and he continues to be under supervision so that he does not aggravate the injured arm.

Another example would be to put a firefighter with a low back injury on fire inspection duty while his back heals. Such patients usually return to full duty much sooner than if they are off work in a typical treatment program.

**Physician's Knowledge of the Workplace**

To facilitate light duty work, the physician must be familiar with the work-sites. There is no alternative to personal knowledge of the workplace. Most physicians find that this visiting of the workplace is a task that cannot be delegated. For physicians who are unable to visit, some success has come from videotaping the job and asking the physician to view it and make comments regarding what the patient can or cannot do.

**Problems with Restricted Duty**

Light duty is not always available for several reasons. Some employers simply will not accept it, based on previous bad experiences with it. Employers in small companies have difficulty providing light duty because they are chronically short staffed. Unions frequently will not allow light duty in their contracts. In some plants light duty is simply not available, and in some the injured worker has to be carried as an "extra" employee; in times of budget cuts this is difficult. In some plants a "light duty" employee cannot cross categories. For example, an injured lathe operator cannot be put to work in the storeroom because it is a different job category. There is considerable patient resistance to light duty in some cases, or even unpleasant peer pressure from other employees.

Still, whenever possible light duty is the best approach when full duty is not possible. Most employers seem to agree that light duty may be difficult, but it is still valuable. Most occupational physicians will agree that returning a patient to restricted duty, and doing it whenever feasible during the rehabilitation period, leads to earlier eventual full recovery

Getting the patient back to work, safely, is the goal of every part of this complicated field of work injury management.

# Section 21

# The Rehabilitation Consultant

*Suzanne Tate-Henderson and Timothy R. Johnson*

## INTRODUCTION

Moving away from "hands on" rehabilitation, the management of the rehabilitation process can be viewed from a broader perspective. Good managers have four basic functions. First, they must be able to plan effectively. Second, they must be able to organize the process and the people involved. Third, they must be able to lead the process from beginning to end. Finally, they must be able to exercise control during the entire process.[1] By reviewing these functions, the rehabilitation consultant can manage better the rehabilitation process until satisfactory case resolution has been reached. Because there are several professions involved in the return-to-work process, and because those professionals have specific functions, it is the rehabilitation consultant who can synthesize, communicate, and facilitate a positive result for all involved.

## THE PLANNING FUNCTION

A plan is a detailed method for the accomplishment of an objective or a task. In this context, the planning function is perhaps the most important of the four functions in that it encompasses both evaluation and rehabilitation plan development. The rehabilitation consultant must be able to gather information, evaluate and integrate that information, set goals and timetables, and identify other roles and functions in the rehabilitation process.

## Gathering Information

For the rehabilitation consultant the information gathering function involves primarily three basic steps[2]:

1. Client interview.
2. Vocational assessment using various standardized tests.
3. Review of all medical facts of the case.

The client interview is important not only as a means to vital client information, but also as a tool to orient the client to the possible steps in the rehabilitation process.

A vocational assessment should consist of measures that are tailored to the individual client rather than the standard battery of tests. Choice of tests should be modified with each individual client, although usually intelligence, achievement, aptitude, interests, and personality testing should be performed. The results of these tests should assist both the rehabilitation consultant and the client to plan the steps in the return-to-work process.

The review of medical information, such as hospital and physicians reports, is important in determining a person's highest vocational potential. Typically, this medical information will accompany the referral. This referral will normally raise critical issues to be studied. The review also will identify medical rehabilitation professionals who may be important team members. The review is combined with other client information to begin determining present and potential transferable skills.

## Evaluating and Integrating Information

The next step in the planning function is the evaluation and integration of all information. The rehabilitation consultant then works with the client to interpret all of this essential information with the goal of establishing a vocational objective. The vocational objective should be designed with the purpose of utilizing all strengths and minimizing all weaknesses on the individual case.

## Setting Goals and Timetables

Once the vocational objective has been determined and agreed on by all parties, the rehabilitation consultant can move to the next phase of the planning function—setting goals and timetables and establishing verifiable steps in the process. Certain laws or jurisdictions may impact on timetables. Also, those same laws or jurisdictions may impact on the nature of goals to be set.

For example, many workers' compensation laws will describe the employer or insurance company's liability with regard to level of employment necessary to terminate benefits. A thorough knowledge of the law is necessary to appropriately establish goals.

### Identifying Other Roles and Functions

Once information has been gathered, evaluated, and integrated and goals and timetables have been set, the rehabilitation consultant and and the client can identify, verify, and communicate all the other roles and functions that will be involved to ensure a smooth working rehabilitation plan.

## THE ORGANIZING FUNCTION

The first step to good organization is to evaluate the various roles and functions of the rehabilitation consultant, which may include:

- rehabilitation counselor
- vocational evaluator
- rehabilitation coordinator
- job-seeking skills teacher
- job placement specialist
- other

All of these functions are not neatly separated by time or space. Therefore, the rehabilitation consultant must prioritize the functions according to the client's needs. Because medical rehabilitation consultants come from varied disciplines (eg, counseling, nursing, occupational therapy, physical therapy) and backgrounds, their approach to a particular case should maximize their strengths. In large consultant firms, a medical specialist may team with a vocational specialist for optimum coverage.

The rehabilitation consultant should have organizational systems available for evaluation, plan development, and placement, as well as community resources and computer applications; all may be integral parts of his or her job. This will facilitate the rehabilitation consultant in viewing the job within the particular setting as an overall framework or skeleton from which to base activities. (See Figure 21-1.) A model of the approach taken in any individual setting is always helpful in allowing the rehabilitation consultant to determine exactly in which phase of the process he or she is currently engaged and the next steps required to ensure the process moves smoothly.

Also, in many states, the consultant will be aided or constrained by law, especially in workers' compensation cases when each of the 50 states has its own unique workers' compensation laws. Mandatory vocational rehabilitation under some state laws will require a finely tuned organizational system to ensure that the client receives the proper benefits.

## THE LEADING FUNCTION

Generally, the leading function involves directing the various components of the rehabilitation plan in pursuit of the vocational objective. If a rehabilitation plan is developed properly, it should provide a step-by-step program that leads the client from present circumstances toward optimal employment. The rehabilitation consultant is now responsible to lead the client and those involved in the rehabilitation process through their responsibilities and functions.

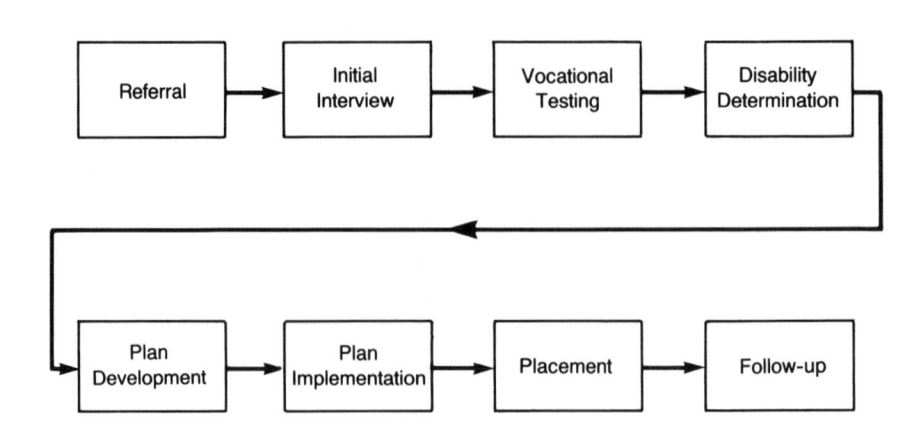

**Figure 21-1** Case management model.

A large part of the implementation of the rehabilitation plan will involve counseling. Clients may need to deal with various emotional barriers preventing return to work, such as fear, acceptance of disability, and acceptance of the vocational alternatives presented to them. Clients also may need to be led through the process of actually obtaining a job and presenting themselves in the best possible light to any prospective employers.

Communication will ensure that other professionals involved in the rehabilitation process will work toward mutual goals in an effective and timely manner.

Obviously, many of the events in a rehabilitation plan and process are interdependent and the successful completion of one component is necessary to the implementation of the next. There must be an ability to anticipate and identify any potential or developing problems in the rehabilitation process so that alternative plans or additional resources can be implemented to minimize those problems. The rehabilitation consultant also must be prepared to refer to other agencies if problems develop that are beyond his or her span of control. Finally, the rehabilitation consultant works with all the participants in the rehabilitation process to facilitate their follow-up on individual roles and responsibilities.

## THE CONTROLLING FUNCTION

Primarily, the rehabilitation consultant is required to influence the people or the steps involved in the rehabilitation process. The rehabilitation consultant must be able to verify or regulate by some form of systematic comparison the steps in the rehabilitation process and those people or resources that are required to implement and complete those steps.

Through the process of regulating and verifying the steps in the rehabilitation process, the rehabilitation consultant will be in a position to continually and actively upgrade the quality of resources available, and to eliminate any ineffective resources.

The rehabilitation consultant must constantly bear in mind that the processes of reporting, especially making recommendations, are the basis on which he or she is judged. If the rehabilitation consultant continually uses resources that prove ineffective, his or her success will also be diminished.

Additionally, the rehabilitation consultants must keep, or have available, statistical data concerning the overall effectiveness of his or her role. This includes such statistics as successful cases, distribution of disability types, time and cost factors, as well as statistics delineating possible causes for unsuccessful rehabilitation ventures. This kind of statistical information will be at its best when the rehabilitation consultant follows up with each case on a regular basis after the completion of the rehabilitation plan. This type of follow-up also will enhance the rehabilitation consultant's perspective of the entire process.

## SUMMARY

For the rehabilitation consultant to function effectively as a manager, each of the four basic components reviewed are essential. Any breakdown in the planning, leading, organizing, or controlling functions can allow a breakdown in the entire process. Although the rehabilitation consultant may work in a variety of settings and under a variety of conditions, this model of coordination and management enables positive results from all available resources through a team approach.

---

**REFERENCES**

1. Allen LA: *Making Managerial Planning More Effective*. New York, Mc-Graw Hill Book Co, 1982.
2. Deutsch PM, Sawyer HW: *A Guide to Rehabilitation*. New York, Matthew Bender, 1985.

# Section 22

# Employer Point of View

*Norma E. Swanbum*

In beginning to discuss a return-to-work philosophy from the employer point of view, a philosophy recommended to employer management by dozens of workers' compensation specialists and physicians for many years—the "CARING" philosophy—should be emphasized. Experience has proved that if the employer has a concern for the health and welfare of his or her employees, not only when injured on the job but as a whole, both employee and management reap the benefits.

It also has been shown that the success of an employer's compensation program depends on the success of an employer's safety program. Concern for employees begins with preventing the injury from happening, follows with maintaining a close relationship with the employee when an injury does occur, and continues with getting the worker back on the job as soon as possible.

## PREVENTING THE INJURY

Management must develop, publish, and distribute to each supervisor and employee a formal policy statement that the company will make every effort to provide safe working conditions, and that safety rules will be enforced. The policy should be written in language supervisors and subordinates understand, should be signed by the chief executive officer, and should form the basis for training programs. Many companies include loss prevention, safety, and workers' compensation costs in salary review for their managers to involve, and place responsibility for safety on, the supervisor.

There is a very close relationship between safety and workers' compensation, and many employers feel strongly that their workers' compensation program should be administered through the safety department. Statistics prove that any employer who has a good safety record also will have a good workers' compensation record. Correlation between these two areas is extremely important for success. The safety department should have the responsibility of administering the safety program and investigating accidents and injuries. Administering the workers' compensation program from the safety department enables personnel responsible for the workers' compensation program to monitor closely all injuries to ensure not only that the worker receives prompt and proper medical care, but that all workers' compensation laws of the state are followed.

## WHEN THE INJURY OCCURS

To develop a successful return-to-work program, the employer must first adopt a caring philosophy and then convince his or her workers that the concern for employees is sincere. The program must begin by dealing with the employees immediately when the on-the-job injury occurs, to assure them that management does in fact "care." Briefly, the employer must:

- obtain the best medical care available for each injured worker;
- communicate with the injured worker and his or her family;
- explain to the employee what is happening (keep him or her informed); and
- be willing to spend a few extra dollars over and above what the state's workers' compensation law requires.

Company policies are involved to carry out an efficient handling of injuries to lay the groundwork for a successful return-to-work program. Although it may be easier for an employer in a large company to develop and carry out some of these details, an employer in a much smaller company also can be successful with some of these ideas. As an employer you should consider developing policies and procedures that address reporting the injury, managing lost time or restricted duty with an injury, and assisting the injured worker with his injury and transition to return to work.

### Reporting the Injury

Develop an in-house first aid report form, which employees can complete to report all injuries, no matter how minor. *Encourage* employees to report all incidents, advising them it is for their own benefit to get the matter on record in case later medical attention should be required, and to prove to management that the injury did occur on the job. Include a section on the form to be completed by the immediate supervisor, to get the supervisor involved. Require the supervisor to inspect the worksite where the injury occurred and to make suggestions on the form as to how that particular injury or another injury of the same nature could be prevented. It is important to get the supervisor involved at the onset, because only with that supervisor's involvement and cooperation can a return-to-work program be successful. (See Exhibit 22-1 for a sample first aid and supervisor's accident report form.)

### Managing Lost Time or Restricted Duty

Develop a company rule on immediate notification to proper management personnel of injuries that require lost time or restricted duty. That policy should include requiring an immediate telephone call to the proper party, advising that such an injury has occurred. The management personnel can then learn the nature of the injury and the name of the treating physician to ensure proper medical care. Then whether the nature of that injury warrants lost time or restricted duty can be determined. The management personnel can then call the attending physician to explain the "restricted duty policy" of the company—that there is light duty work available, if the employee's medical condition permits. A company can prevent many injuries from becoming lost time injuries with this policy because most medical providers do not realize that a company would have restricted duty available. When the injured worker reports for medical attention and advises that physician of the nature of his or her job duties, often a specific injury would not permit the employee to return to regular duties. But, if the physician knows restricted duty is available, he or she will often release the worker to perform lighter duties until he or she is healed.

### Supporting the Injured Worker

Develop a policy whereby management personnel or the immediate supervisor is on hand immediately after the injury. Send someone to the hospital with the ambulance, or with the employee when he or she goes for emergency medical treatment. Not only does this reassure the injured

**Exhibit 22-1** Sample First Aid and Supervisor's Accident Report

---

Form No. 4034 6/84

**MINNESOTA POWER**
**FIRST AID AND FOREMAN'S ACCIDENT REPORT**

NOTE: THIS FORM IS TO BE COMPLETED IN ALL TYPES OF INJURIES, WHETHER OR NOT MEDICAL ATTENTION IS REQUIRED. IF MEDICAL ATTENTION IS REQUIRED, PLEASE ALSO COMPLETE THE "FIRST REPORT OF INJURY".

AREA OFFICE, GENERATING STATION OR DEPARTMENT: _____

| DIVISION OR DEPARTMENT (check one) | REGION SAFETY COMMITTEE (check one) |
|---|---|
| ☐ CENTRAL | ☐ CENTRAL REGION |
| ☐ NORTHERN | ☐ NORTHERN REGION |
| ☐ WESTERN | ☐ WESTERN REGION |
| ☐ POWER SUPPLY | ☐ POWER SUPPLY |
| ☐ SYSTEM ENGINEERING & OPERATIONS | |
| ☐ GENERAL OFFICE | |

Date of incident: _____  Time: _____ AM PM

(1) Name of injured (print): _____  Occupation: _____

(2) How long in department? _____  Place where injury occurred: _____

(3) Injured's immediate supervisor (print): _____  Was he present at time of incident? _____

(4) Had employee been instructed in the work? _____  By whom? _____

(5) Machinery, tool, or equipment involved: _____

(6) Is it in good condition? _____  Is a safeguard provided? _____

(7) Describe injury or illness in detail and indicate part of body affected: _____

(8) Describe in full how incident happened: _____

(9) Witnesses: _____

(10) Was injured wearing: Safety Glasses? _____  Hard Hat? _____  Safety Shoes? _____

(11) Other protective equipment? _____  (What)? _____

(12) What was the principle cause of the incident? _____

**Exhibit 22-1** continued

(PLEASE ANSWER QUESTIONS FULLY)

(13) Were there safety rules or measures in effect to prevent injury? _____

(14) If so, describe: _____

(15) Would additional safety measures have prevented injury? _____ If so, describe: _____

(16) What do the injured employee and supervisor suggest to prevent a repetition of such an injury? _____

(17) Should this injury require medical attention, notify immediate supervisor at once. (Boswell plant employees call Extension 4615.) I certify that I have read line (17) and will comply upon receiving medical attention.

**TO BE FILLED OUT BY SUPERVISOR:**

(18) Did injured have professional medical attention? (Yes or No) _____

If yes, where? _____

(19) Was first aid administered? _____ How long after injury? _____

(20) Will injury result in lost time other than on day of injury? (Yes or No) _____

**CORRECTIVE ACTION:**

(21) Incident site inspected by supervisor?  ☐ No  ☐ Yes  (Supervisor initial and date): _____

(22) Corrective action taken (explain): _____

(23) Work order issued to correct the problem?  ☐ No  ☐ Yes  (Work Order number): _____

(24) Supervisor's comment (If no immediate action can be taken, recommend how to prevent reoccurrence): _____

(25) Do injured and supervisor agree? _____

_____  _____  _____  _____
Injured Employees Signature  Date  Supervisor's Signature  Date

*Source:* Courtesy of Minnesota Power, Duluth, Minnesota.

worker that he or she will receive proper care, but also it can be of significant assistance to a spouse or family member who also reports to the treating medical facility to be with the patient.

Have someone from management or an immediate supervisor visit the injured worker in the hospital or at home to explain the workers' compensation benefits, to advise him or her who will pay the medical bills, and to ask if anyone in the company can be of assistance with anything for the worker or his or her family while the employee is disabled. Again, the employee knows the employer "cares."

While an employee is off work, have someone from management keep regular contact with him or her (at least every 2 weeks). Personal contact at the residence is best; but if that is not possible, at least regular telephone contact should be maintained.

All of these policies and procedures will be of extreme assistance when the the return-to-work policy is put to use.

## RESTRICTED DUTY PROGRAM

How can you, the employer, develop a strong return-to-work policy? First and foremost, initiate a "restricted duty" program. Don't expect an injured worker to be able to return to full duty immediately unless you are prepared to suffer the consequences of expensive workers' compensation costs and the possibility of longer than necessary disability. When employees are off work for an extended period of time they develop several problems other than the injury itself—psychological problems, marital problems, or family problems. In the best interest of everyone concerned, attempt to get the employee back to work as soon as possible.

After the restricted duty program has been established, the employer must begin communicating with the attending physician and the treating physical therapist or occupational therapist. Contact the medical provider to obtain specific work restrictions, and if possible, get those restrictions in writing.

A new policy related to restricted duty was developed recently at the company where the author is employed, and it has already proved to be most successful. After obtaining the work restrictions from the medical providers, a meeting (to be held at the injured worker's jobsite) is arranged to include the supervisor, the injured worker, and the workers' compensation administrator. At that time, the restrictions provided by the physician or therapist are shared with all parties (at the same time) to ensure complete understanding by all concerned. The employee is reminded that the physical restrictions apply not only while on the job but also in off-the-job activities. Holding the meeting at the jobsite provides an opportunity for on-site examination of the restricted duty that will be available to the employee to assure that those duties meet the restrictions placed by the medical provider. Not only does this policy get the supervisor involved in the situation, but the injured worker again knows you "care."

## COMMUNICATING WITH MEDICAL PROVIDERS

Communicating with medical providers is one of the most important factors in any return-to-work program. Don't be afraid to call the attending physician, physical therapist or occupational therapist, or chiropractor. These people don't know your company policy—make them aware of it. Provide the medical person involved with the written job description for the injured worker and any other information available that will assist him or her in making a determination as to return to work. Explain in detail the restricted duty that would be available to the employee; don't just tell the medical person you "have light duty available." Explain your concern for the health and welfare of the employee; and *bear in mind* the job of the physician or therapist is to treat patients, not to be automatically aware of your company's policy or the workers' compensation laws in your state. Oftentimes medical providers tend to believe the employer is an adversary because the patient was injured while on the job. Make them aware of the fact that you are there also to assist with the healing of the patient.

Remember, the medical provider probably has no idea as to the requirements of jobs within your particular company, or even the physical structure of the worksites at your facility. Invite the physician or therapist to visit your worksite so that he or she may personally view the working conditions and jobs within your company.

One of the injuries at the company where the author is employed involved a knee injury with resulting surgical knee reconstruction for an employee working in an electric generating station. Because of the surgery, the employee will be required to wear a full and very restricting knee brace for the remainder of his life. When an attempt was made to obtain information from the surgeon as to whether or not this man would be able to return to his preinjury occupation, even though a written job description and physical requirements of the job were provided to the physician, the work restrictions received included no information as to climbing restrictions. The plant in which the employee worked contains overhead open steel grating and open grating on stairways, and his job required considerable scaffold and ladder climbing. (See Figure 22-1.)

324   Work Injury

**Figure 22-1**   Generating station stairway.

---

The employee returned to work on a restricted duty basis, but the knee brace ordered by the surgeon had not yet arrived. On a weekend, at home, the employee fell down a flight of his basement steps because his injured knee "gave out," and the employee suffered compression fractures of two thoracic vertebrae. Because the fall was caused by the knee that was injured on the job, the company had to assume the workers' compensation responsibilities for the vertebral injury. To prevent further injury and to obtain information as to whether or not the employee should be working in that plant (even with the restricting knee brace), the company retained the services of an occupational medicine specialist. The specialist visited the plant, reviewed the job at the site, and then examined the injured worker. The evaluation of the jobsite and the employee resulted in information indicating that the employee should not work in his preinjury position or at that facility, with his permanent knee condition. The result was that another job was found within the company that would accommodate the knee brace and restrictions.

This particular case demonstrates the importance of communicating with physicians and therapists to obtain their cooperation in providing necessary medical information for your worksite and specific jobs within your organization.

## MODIFYING A JOB

Several questions have been posed as to whether or not it is worthwhile for an employer to modify a job during the injured employee's healing period and/or to accommodate a permanent physical problem that is a result of injury. It has been established that if the employee isn't working, workers' compensation costs increase and the employee can develop problems other than those caused by the injury. He or she can become accustomed to staying home, making it more difficult to return the employee to work. It has been shown that both the employer and the employee benefit when the injured worker stays in contact with the rest of the work force. This not only keeps the injured worker's mind occupied, but provides a continuation of knowledge that the employee has gained in his or her work experience so he or she will be more productive on return to regular duties. If the injured worker can do "anything" it is better than doing nothing by staying home. In addition, any meaningful work that can be performed results in better production for the employer.

If the employer has no one on staff with expertise in modifying jobs, there are ergonomics specialists or therapists available in most areas of the United States to assist with these projects. Many times simply modifying the height of a workbench, raising or lowering of chairs and stools, or providing mechanical lifting devices can solve a job modification problem.

## PEER PRESSURE FROM CO-WORKERS

What can be done to avoid peer pressure from other employees when an employee is working on restricted duty or in a modified job? Management should inform all co-workers of the physical restrictions of the injured worker and remind the peers that it could be them someday who would need cooperation and assistance on the job. Explain to the peers (as much as confidentiality rules permit) what the injured worker is going through, that he or she wants to work, that he or she may need assistance, etc. Communication is the key to the success of any return-to-work program. If the co-workers also understand the "caring" philosophy of management, they are more likely to "care" about the injured worker.

## POSSIBLE PEER PRESSURE FROM THE SUPERVISOR

What can be done to prevent peer pressure from the supervisor? Earlier the philosophy was presented that it is only with the supervisor's involvement and cooperation that a return-to-work program can be successful. The success of the program, as far as the immediate supervisor is concerned, again becomes a matter of communication. Management should educate its supervisors in the company's philosophy. The supervisors also must learn to "care."

The supervisor is the key person in on-the-job injuries and must be trained to handle those injuries accordingly. The following are some suggestions in that regard:

- The supervisor should know the facts surrounding the incident and be involved in the investigation. The basic elements in the investigation are evidence, witnesses, and determining the cause of the incident.
- The supervisor should know the nature of the injury and how long any resulting disability or restricted duty is likely to last.
- The supervisor should learn from management personnel who will be handling the case, whether or not physical restrictions will be placed on the employee on his or her return to work.

To accomplish these goals the supervisor should be trained

- to make sure the injured worker seeks immediate medical attention
- to document the incident on the forms that management has developed and report it through proper channels within the company
- to maintain regular and personal contact with the injured worker

Supervisors should be encouraged to visit their employees at home or in the hospital, telephone them, and offer assistance, whether or not the absence is work related. The supervisors should be trained as to the importance of the employee being able to work on restricted duty, both for the advantage of the company and for the benefit of the employee. They must understand the possible inability of the worker to work at full speed. If the worker has problems, the supervisor should check the source of those problems and deal with them accordingly. The supervisor must realize that the injured employee may not be feeling "up to par." Management should not penalize a supervisor for low productivity when a disabled worker returns to work.

Supervisors can be educated to become involved with an injured worker and can become an important asset in the success of a return-to-work program.

As another case study, one of the employees at the company where the author is employed was scheduled to be off work for a considerable period because of the aggravation of an on-the-job back injury. When injured, the employee had been a "physical worker," performing heavy construction work. After the injury and resulting surgeries, he was placed in a lighter duty full-time position that involved computer terminal entries and programming. The aggravation occurred after he was working in the lighter duty position. The immediate supervisor, having been trained to "care," took the initiative to deliver a computer terminal to the employee's home and have it connected to the company's computer system so that the employee could perform at least a portion of his regular computer duties from his home while recuperating. The efforts of the supervisor resulted in the disabled employee keeping occupied; and even though he was performing only part time, he was doing productive work for the company and his department.

## LABOR UNIONS

Labor unions also can play an important role in the return-to-work or job transfer process. The employer must have rapport with the bargaining unit or units that represent his or her employees. Whether or not a labor union will be cooperative in a return-to-work situation cannot be predicted. However, the employer can do everything possible to attempt to develop cooperation. Management should educate the union's business agent. Explain what you can about workers' compensation and how it is handled by your company or insurance company. Also, explain the company's policies regarding, for example, return to work and job transfers, and make the union aware of the fact that you do "care" about the workers and do want them to return to productive work. Offer assistance to the business agent with questions that union members may pose to the union as to workers' compensation or company policy. If the union managers know that the employer is willing to be cooperative and assist with injured workers, they will be more willing to cooperate with management.

## COMMUNICATION

Communication among management, labor unions, and workers is the key word in dealing with on-the-job injuries. By communicating with one another these groups can carry out a "caring" philosophy.

In the company where the author is employed there is a continuing policy to communicate management's philosophy and company policies to the workers. At least once each year, at one of the monthly safety meetings held by each department in the company, the workers' compensation administrator conducts an informal session to advise workers of the details of the state's workers' compensation law, and the legal rights of those workers should they be injured on the job. The administrator explains what happens "if" workers would be injured, and in this manner the company develops a rapport with the workers so they will know who they can contact if they have a question or a problem. Management feels that these meetings, which include emphasizing the caring philosophy of the company, tend to keep the employee "on the company's side" after the injury. The workers may not remember the details of what they have heard at the meeting until they suffer an injury themselves because most people have the usual attitude of, "I never thought it would happen to me." But, they will remember that management is willing to do everything possible for them when they are injured, and they will learn that the company does "care."

Whether dealing with the management or operation of a company, on-the-job injuries, or a return-to-work program, no program can be successful without communications. Everyone must communicate with one another, from top management on down through the ranks to have a successful business or a successful return-to-work program.

## INSURERS

The purpose of the workers' compensation insurance carrier is to handle injury reports after they have been filed by the employer, to file necessary papers with the governing state agency, to make benefits payments to the injured worker and medical payments to the health care providers, and to act as consultants to the employer in the workers' compensation law and its interpretation. However, the insurer cannot manage a workers' compensation claim for the employer.

Unfortunately, too many employers consider the workers' compensation insurance carrier as they would their automobile insurance carrier: "We pay the premium—so you manage our claim." The insurance carrier can be of assistance in many matters involving an on-the-job injury, including providing advice in the development of return-to-work programs. However, the insurance carrier does not know the employees of the insured employer and is not familiar with the worksites and jobs of the employer. Only the employer and management knows its workers and can successfully manage the claim of an injured employee.

Specialists in the workers' compensation field give excellent advice to employers; therefore you should consider the following tips:

- Become involved with the insurance carrier.
- Insist that the carrier make prompt and proper payments of benefits to the injured worker.
- Advise the carrier of your company's philosophy and procedures.

**DON'T**
- Sit back and let the carrier deal with the injured worker.

The policies and philosophy that have been established to deal with injured employees must be maintained by the employer and its supervisors!

## SUMMARY

Some recommendations as to how employers can have a successful return-to-work program are the following:

- Develop a "caring" philosophy with your workers.
- Be sure the injured worker gets the best medical care available.
- Keep the employee informed of work injury processes.
- Maintain regular contact and communications with the injured employee.
- Educate the supervisors and keep them "involved."
- Don't be reluctant to spend a few extra dollars—it pays in the long run.

Remember the "caring philosophy"—CARING ABOUT PEOPLE. A closing quote from the brochure entitled, *Workers' Compensation*

*System—A Better Way To Manage It,* published by the Minnesota Business Partnership, tells the whole story[2]:

Workers' compensation was meant to be a simple system, providing fair compensation for job-related injuries while fostering good employer/employee relations. Yet the system has strayed from its original purpose and has become costly and complex.

Who is to blame? WE ALL ARE. Government must do its part, but so must we. Every employer, regardless of size or type of business, can do something about cutting costs of injuries—it's a matter of attitude.

We need to CARE enough.

We CARE by providing safe working conditions. We CARE by taking a direct interest in employees injured at work. We CARE by being creative and by finding available resources to solve problems.

If we are committed to these principles we can take a big step toward better management of workers' compensation.

---

*Reprinted from Workers' Compensation System—A Better Way To Manage It (p 14) with permission of Minnesota Business Partnership.

# Section 23

# The Therapist

*Susan J. Isernhagen*

Roles of the physical therapist and occupational therapist specializing in the occupational medicine field are expanding.

Physical therapists have historically been involved in care of the work injury. Although necessary for musculoskeletal injury, treatment is now seen as only the first step in progressive management of the injured worker. The physical therapist can use skill in evaluation and functional restoration for a more comprehensive approach. With a background in functional living skills, the occupational therapist adds depth to the program once the patient is stabilized. Creativity in work simulation is a growing role for progressive departments.

The team of physical therapy and occupational therapy can have a powerful effect on work injury management when both are available and work together toward common goals.

The therapist's work injury management role is clarified if viewed as a "rehabilitation" therapist rather than as an acute care practitioner. For example, the diagnosis of a patient with a stroke or head injury is accepted as fact. Treatment goals are not to heal the brain but rather to increase function to maximum capabilities.

The work-injured patient should have the benefit of similar goals. It is important to treat the injured part in regard to pain relief and restoration of function. Once beyond the acute or intermediate level, however, life style also should be considered. In the occupational medicine patient, return to work will necessitate not only healing but restoration of work function by the patient as a whole.

For example, a patient's "healed" back injury may allow relatively pain-free movement, but is the worker ready to resume lifting, carrying, prolonged sitting, bending, or overhead work? Have the quadriceps become weakened and has aerobic capacity diminished? Will the injured part or any other deconditioned functional area become an injury-producing liability if work is resumed?

The therapist's attention must then concern functional evaluation and a safe return to work. Just as the rehabilitation therapist intervenes with the patient until return to home is possible, the occupational medicine therapist works through the return-to-work process.

There are defined attributes and philosophies that prepare the therapist for a strong role in work injury management.

## AN EXTENSION BEYOND THE MEDICAL MODEL

Work injury management requires a basic understanding of the workers' compensation system. Each professional has a role, and to interact properly, a knowledgeable therapist will know the professionals involved in the process. Employers are outside the medical model; yet, they have a significant need to interact with the medical caretakers of the employee. Because medicine is a science, with specific jargon and methods, care should be taken to bridge the gap between medicine and industry. Willingness to communicate and explain are attributes of the therapist in the occupational medicine field.

## ACTIVE COMMUNICATION

Once functional communication is accepted as a goal by the therapist, active efforts to facilitate the exchange of information can be planned. Proper releases of information should be obtained by the therapist to ensure dissemination of appropriate reports to employer, insurance company, and workers' compensation representatives. This prevents adversarial attitudes that can arise when one party knows something the other does not. It also reduces misinterpretation of the patient's physical findings, as all involved, including the worker, receive the same report.

The physical therapist and occupational therapist may be part of a multidisciplinary team. Institutional teams should present a common report, even though each discipline may have unique contributions.

## OBJECTIVE AND UNBIASED POSITION

Traditionally, medical practitioners have been patient advocates. To the extent that a patient's needs are fully met and they receive positive support in their quest for health or function, this role is positive. In a situation where the medical healing is complicated or there is the possibility of outside influences, however, being objective and goal oriented is more beneficial to the functional process. The therapist that can work toward medically and physically oriented goals for the patient's benefit finds it much easier to retain a professional attitude than one caught in the middle of opinion differences.

For example, a patient-worker may be slowly recovering function in a work hardening program and yet be reluctant to return to work under a punitive foreman. A natural patient-oriented feeling that may arise in the therapist might be sympathy for the patient-worker and a reluctance to push the work return. The therapist can take a more helpful approach, however, by increasing work function and then communicating ability and limitation recommendations to appropriate professionals. Often employers and supervisory personnel will accept job restrictions if the physical reasons for the restrictions are clearly delineated. In addition, empowering the employee to control the work level within those guidelines will facilitate accurate self-judgment of work activities. Both "sides" benefit from clear explanation and rationale of work function by the therapist.

## TRANSFERABLE SKILLS

Because occupational medicine programs involve transferable skills based on educational preparation and experience, the therapist that has ability in functional capacities evaluation will find work hardening a logical extension of those skills. By refining these skills, further ergonomic or prevention programs can be developed from this base.

When therapists who are involved in management of work injuries find themselves asked to go further in program development, optimism can be tempered with thoughtful analysis of new program needs. The therapist should judge whether the new areas are actually extensions of professional skills. If they are appropriate, development will have a solid foundation. If not, there should be a referral to the correct profession.

## FOLLOW-UP/CLINICAL RESEARCH

Because workers' compensation costs are escalating rapidly, programs that reduce costs are in demand. Whether these programs are management based (recordkeeping, incentive and reward systems), education based (prevention programs), or medically based (physical pre-employment screening and more effective injury management) is immaterial. Business will participate in systems that save money.

From a medical vantage point, the practitioner is interested in less human suffering and effective medical recovery systems. This coincides well with the financial incentives, because reduced work injury or reduced worker dysfunction is a goal of both industry and medicine. Areas of interest are:

- prevention of injury—measured in reduced injuries and reduced cost of injuries
- timely and effective treatment—measured in reduced lost time from work, reentry at former job, and reduced cost of treatment
- prevention of reinjury—measured in reduced second injuries, reduced costs, and increased productivity
- cost of functional assessments or work hardening when compared to costs of lost time medical or disability costs of a similar population not treated or evaluated
- functional level of injured employees who have gone through a restorative program compared to those who have not
- expected length of stay in a restoration program for different diagnostic groups
- comparison of worker variables (age, gender, size, health habits, type of work) with performance or injury rates
- establishment of functional norms for workers' in various types of work activities

## THERAPIST ROLES IN RETURN-TO-WORK PROCESS

Depending on the level of the occupational medicine department's status, the therapist may have several roles in the return-to-work process.

### Immediate Care

The physical therapist may see the worker-patient in an immediate/emergency situation if the therapist has direct access to an occupational health department in industry, an emergency department, or a clinic that specializes in occupational medicine. (See Section 11, *supra*.)

The focus of the therapist's role in immediate care of work injury is twofold:

1. Effective short-term care that stresses treatment and protection of the injured part, but does not interfere with use of uninjured parts.

2. Instruction to the patient regarding self-care of the injury, including modalities he or she can apply at work or at home.

## Intermediate Care

If residual injury symptoms continue, modalities and protection would be joined by restorative exercises.

Further education to review the mechanism of injury so that it does not reoccur and to discuss alternative work methods from ergonomic and functional ability perspectives is beneficial.

Occupational therapy to incorporate functional work tasks is a complementary addition to restorative physical therapy at this point.

## Extensive Treatment

In more severe injuries, physical and psychological retraining will begin. With medical approval, work hardening principles can now be implemented. Specific work functions should be used. The physical therapist and/or occupational therapist also should facilitate motivation and de-emphasize disability. Realism is necessary, but the therapist must not discount the recuperative powers of the motivated.

In June 1987 Bart Connors, an Olympic gold medal gymnast, told an American Physical Therapy Association audience that only a few months after a severe shoulder injury and reconstruction he was told by some medical professionals that he would most likely not recover enough function to ever compete as a gymnast again. Other medical professionals were not discouraged and worked with him to gain as much function as possible. Months later, Bart won a gold medal in the Olympics in gymnastics.

His point is well taken in the occupational medicine realm. Physicians and therapists who work with injured employees can have a powerful effect, either negatively by implying "you'll never work at that job again," or positively by stating "we'll work toward your best functional level and a successful return to work."

## Functional Capacities Testing

Section 14, *supra*, describes a comprehensive functional capacity testing designed to describe abilities and limitations to facilitate a safe return to work. This is helpful for all patients in extended treatment. Even for workers in immediate care or intermediate care the therapist should take the responsibility of describing work function when discharging a patient. Any gradation from individual task testing to full functional assessment can be utilized.

**Work Hardening**

After proper treatment and a functional capacity evaluation, a work hardening program might be indicated. (See Section 15, *supra*.) This systemized work rehabilitation program blends the core philosophies of physical therapy and occupational therapy with the needs of industry. Both employee and employer benefit when a stronger functional worker returns to a job that matches the physical abilities present. The therapist is able to assist in improvement and development of skills, make recommendations for future functional (work) performances, and participate as a team member in the actual return to work.

**Conferences**

In an extended or severe injury situation—cases that would require a functional capacities evaluation or work hardening—a conference is the most effective way to disseminate results and recommendations. By taking the conference initiative, the therapist prevents reports from being delayed or misplaced, prevents misinterpretation of findings—either accidental or purposeful—and allows recipients of the information an opportunity to ask questions for clarification and future planning. The therapist who arranges the conference takes the lead in explaining the results of testing or treatment. The reports are read and discussed in detail, making sure that all language is understandable and implications are clear.

Those most valuable to be invited to participate in the conferences are the patient-employee, the employer, the physician, and the rehabilitation consultant or insurance representative. Others who may be interested in selected cases would be the employee's "significant other," attorneys, or union representatives.

Objectives of the conference, which should be stated initially, would be:

- dissemination of physical return-to-work information
- direct discussion of physical abilities and limitations pertaining to work
- initiation of planning the actual physical return-to-work parameters.

All aspects of the conference should be documented and this record would be added as an addendum to the discussed reports.

## SUMMARY

The therapist's role in the return-to-work process can be characterized by return-to-work goal setting, accurate and descriptive functional work evaluation, and physical restoration that is directed at productive work rather than body perfection.

The therapist accomplishes these objectives through interest and expertise in work injury management programs, by communicating both within and outside of the medical realm, and by taking steps to ensure and document a positive work result.

# Section 24

# The Attorney's Role

*Susan J. Isernhagen*

## INTRODUCTION

Attorneys are not involved in all work-injury cases, but when they are, professionals in rehabilitation and work injury management need to be aware of their viewpoint. There also is possible involvement at a future date in any particular case.

The employee most often seeks an attorney when there is a fear that fairness will not prevail in the management of his or her work injury. This is more likely to occur when there is already mistrust between employer and employees. Past history involving resolution of work injury cases leaves a positive or negative impression that future injured employees remember. The negative outcomes lead to higher retention of attorneys by injured employees. Severe injuries also are more likely to necessitate legal intervention.

Professionals (medical and vocational) working with the injured employee are concerned with recovery from injury and safe, effective return to work. The actual or potential involvement of litigation may sidetrack the proceedings and confuse the issues and directions. Both plaintiff (employee) and defendant (employer) contribute to this as each attorney looks at treatment, evaluation, and case resolution differently. One person's situation may look like two completely different cases when viewed through the selective eyes of the two opposing attorneys.

To get a broader perspective of the attorney's role in the return-to-work or case resolution process, three experienced attorneys were asked their viewpoints on litigation and their own role in workers' compensation cases. There are sharp contrasts in points of view, yet the review of these opinions provides a clearer understanding of why attorneys react as they do.

## VIEWS ON THE WORKERS' COMPENSATION SYSTEM AND LEGAL PROCESSES

### James Courtney, III—Plaintiff's (Employee's) Viewpoint:

In pursuing complicated legal theories, the easiest way to get to the bottom of any transaction is to "follow the dollars." It is important to follow the dollars of the transaction rather than the notion that what we are seeking to do is that which is "best for the patient." It is commonly said that what is best for the patient is a return to work. What is not said is that returning the patient to work is also the cheapest outcome for the insurer.

The economics of the situation tend to make the lawyer for the claimant work for the economic best interest of the claimant, since, because of the contingent fee, the lawyer gets paid only when the claimant gets paid.

### William D. Sommerness—Medical-Legal System Viewpoint:

I do not think the workers' compensation system works well if the theory is to get the employee back to work with minimal damage to employer and employee. Although it was originally viewed that much was gained from the founding philosophy that the employee does not have to prove the employer's "negligence" in setting up a safe place to work, the employee simultaneously gave up the right to "hit the jackpot" for any particular injury. Because all monetary rewards are on a schedule, the system simply no longer works as it was designed. The paper work and bureaucracy are almost prohibitive. The rule of procedure, the substantive procedures, all seem to work to everyone's chagrin.

As it stands now, it appears that even if a person wishes to get the case settled, the client wants to get back to work, and the employer is amenable to their situation, the red tape of the "system" may preclude rational approaches to the solutions.

### Ronald Fischer—Defendant's (Employer's) Viewpoint:

Generally, employers perceive that the workers' compensation system is not fair or impartial in any objective or aesthetic sense. Workers' compensation costs are driven upward by awards to employees for injuries and disabilities that employers often view suspiciously. Although the employee technically bears the burden of proving his claim, the burden is often met with minimal evidence. The employer is often in poor position to contest a claim; for example, when huge benefits are claimed by an employee for an unwitnessed injury. The employer may be able to prove that the employee suffered from a significant disability before the injury, but it is most often to no avail as the employee need only allege that the work incident aggravated his pre-existing condition. Very often the finding of a work-related aggravation will be based on the employee's subjective complaints. The employee may only, for example, testify that his back hurt much more following the work injury. Objective evidence corroborating the employee's complaints of greater pain after the work incident has not been required.

Historically the system is skewed to the disadvantage of the employer. The workers' compensation system was devised to dispense necessary subsistence benefits to injured employees. Throughout much of the history of the workers' compensation system in Minnesota, the benefits to employees were very minimal. For example, up until 1967, employees would receive no more than $45 per week for total disability. Lump sum awards for residual impairments were not paid until the employee's return to work. Significant and formative case law interpreting and applying the Workers' Compensation Act, much of which pertains to present disputes, was decided over that period of time when the employee was viewed with especial sympathy. Awards to the employee in a questionable case would have helped an employee to subsist without necessarily imposing a huge financial burden on his employer. Also, judges often justified awards of benefits to injured employees on reference to a principle (recently statutorily eradicated) that workers' compensation is remedial in nature and is therefore to be construed liberally in favor of the employee.

Unfortunately for the employer, the stakes have increased considerably over recent years. Presently (1987), disabled employees may receive up to $376 per week in tax-free dollars. Additionally, employees often qualify for large lump sum payments for residual impairments amounting to many thousands of dollars. Large potential benefit awards, when coupled with the uncertainties of employment in large Minnesota industries, create huge incentives to malinger or exaggerate injuries. The workers' compensation system now imposes huge costs on Minnesota employers.

As indicated above, the employer is often in a very poor position to defend against a suspicious claim; for example when the claimed injury was unwitnessed. Also, given human nature, workers' compensation judges, regardless of background, will more readily identify with injured employees rather than a large insurer or faceless employer. Simply put, all else being equal, a judge will feel better about awarding benefits to an employee and his family than in denying such benefits. The result is that employers are presently burdened with unreasonably high workers' compensation costs.

# PERSPECTIVES ON THE ATTORNEY'S ROLE REGARDING A CLIENT

**James Courtney, III:**

The legal community is essentially a financial physician. The lawyer for the employee is attempting to make the best economic situation for the client.

In my practice I often see employees who feel as though their needs and wants are the last consideration to be met in a return-to-work situation. Very few clients are actually malingerers or people intentionally cheating on the system. Each client, however, is a person attempting to minimize his loss or maximize his gain just as he did before he was injured.

By the time an employee has been off work for a year, he frequently has had his workers' compensation checks interrupted on two or three occasions. His income has been substantially reduced despite the compensation. He has been dunned by medical vendors seeking payment of bills that have been refused by the workers' compensation insurer. He has been called a deadbeat, an exaggerator, and has been told "it's all in your head." He has been given the impression that the next compensation check may or may not come depending on his compliance with the dictates of the rehabilitation consultant who talks in terms of "the injured worker's best interest," but is paid by the insurer, sent to him by the insurer, and secretly reports to the insurer. It is understandable that the client comes to the rehabilitation process somewhat defensive and suspicious. Because of this, my clients frequently feel that the plaintiff's attorney is the only member of the cast that is "on their side."

**William D. Sommerness:**

The role of the plaintiff's attorney is to get the claim filed and the paper work process begun so as to get the other professionals "doing their thing."

Make sure there is a complete and adequate work history of the client and a complete and accurate health history of the client. It can be devastating after a number of months of work to have the attorneys for the defense come up with previous injuries and illnesses that point to a pre-existing condition that the client never told you about. Candid analysis and appraisal of the case after the gathering of the health and work history is the most critical part of any case.

**Ronald Fischer:**

In representing an employer-client, I investigate and evaluate the employee's claims. Observations of the employee's ability to function outside the medical examining room may shed light on complaints of liability. A thorough functional capacity evaluation may aid in evaluating subjective complaints. Very often it is the lawyer's responsibility following an investigation to advise his client to pay the questioned claim.

In cross-examining treating physicians, employer's counsel may attempt through preparation and nonhostile demeanor to influence the witness's opinion more fairly to the employer's position. I select independent medical examiners whose credentials will reduce the likelihood of suspicion that they are "hired guns." As a defense advocate in a workers' compensation proceeding, my role is to fairly represent the employer and to hold the employee's case to its burden of proof.

## WHAT IS THE ATTORNEY'S ROLE IN RETURN TO WORK?

**James Courtney, III:**

When the employee does return to work, he often feels that his needs and wants are the last consideration to be made in his return-to-work process. He is treated by fellow workers and foremen as a malingerer. Co-workers resent accommodations made for the injured worker. Employers resent the injured worker's limitations. Frequently the light-duty jobs are make-work jobs giving employees no pride in life or job.

Once an employee has a work-related disability, he becomes factually unemployable whether or not he theoretically could be employable according to his residual physical capacity. This is the concept of the "odd lot" disability. Employers faced with the pool of potential employees will pick the strongest. The odd man is out.

In my practice I see time and time again employees who feel that their needs and wants are the last consideration to be met in a return-to-work situation. We all work for money, but what makes each of us tick is the feeling that we are doing something constructive and worthwhile. When an employee sustains an injury that does not heal within the socially accepted period, he becomes a leper at the jobsite.

What disables 80% of the people from work is pain. It is not a physical inability to reach, stretch, or bend. Rather, it is the pain attendant thereto. Pain is perceived differently by all of us. What evaluation cannot do is define the amount of pain a patient should be required to tolerate on a day in, day out basis.

**William D. Sommerness:**

My advice to attorneys thinking about participating in the return-to-work process is "stay out of it." Although that may seem a little facetious, I think the attorney should let the professionals in the other disciplines "do their thing." I sometimes sense the other professionals hold back on what they really think if the attorney is a little too active, too demanding, too intrusive into their sphere. They are concerned that if they don't do what the attorney thinks, they may be subject to criticism at the time they testify. That may or may not be the case, but if the idea is to get the employee back to work in as functional capacity as we can, in as reasonable a time period as we can, then I think the attorney ought to get on the sidelines and let the people who are trained to do that do that. On the other hand, sometimes the attorney has to get involved because the medical team seemingly only responds to those who are the loudest about the needs of their particular client-patient.

**Ronald Fischer:**

Except for the malingerer (in my opinion, an increasingly greater phenomenon with today's higher benefit structure), the injured employee and his employer share a mutual interest in the employee's returning to work as soon as medically advisable after injury. An early return to physical activity is therapeutic. A study was conducted by the University of Texas Health Center in San Antonio. Patients at a walk-in clinic who complained of low back pain symptoms were randomly prescribed either of two treatment modes. One half of the patients were prescribed two days of bed rest, whereas the remainder were directed to remain in bed for seven days. The patients were contacted following their eventual release from treatment. Those members of the group who had been prescribed the shorter bed rest experienced 45% fewer lost days at work. Also, those patients who had been prescribed the longer period of bed rest were more dissatisfied with their medical care and were more likely to have changed treatment to other medical care providers.

In attempting to fairly reduce my client's workers' compensation costs, whenever medically advisable I promote an early return to work. This may involve educating a reluctant client to the mutual benefits of an early return to work. Work hardening programs should also be considered. An early return to work will generally reduce the likelihood of a long period of claimed disability that would in turn give rise to an adversarial situation between the employer and the employee. At the early stages of recovery following injury, by a show of genuine concern for the injured employee, an employer may reduce the likelihood of costly and unnecessary litigation.

## HOW DO YOU VIEW THE ROLE OF OTHER PROFESSIONALS IN THE WORKERS' COMPENSATION SYSTEM?

### James Courtney, III:

The economics of the situation get in the way. Although the rehabilitation consultants are supposed to remain neutral, they are, in fact, paid by the insurers, and their very existence depends on continued referrals. It follows that in the event they do not act in the best interest of the insurer, they will not be hired on the next case.

We are seeing an increasing use of industrial medicine physicians in the workers' compensation field. The increasing practice is to have the employer's physician take over the "treatment" of the employee and make the determinations about the employee's compensation benefits. We are seeing increased circumstances where the industrial medicine physician is the judge, jury, executioner, and keeper of the flame. This multifaceted physician's role is marvelous for all concerned except the employee. The industrial medicine physician tells the patient that he or she is "only doing what's best for you (the patient)" while regularly determining that "what's best" is that which his employer wishes. This situation increases the feeling of helplessness and powerlessness of the employee.

Even his physician is no longer "his."

Many clients have also asked whether, because their functional capacities evaluation says that they can work eight hours a day, they are now required to spend their weekends and evenings on the couch, on a heating pad, giving up their entire social life to comply with the return-to-work recommendation of their functional evaluation and the rehabilitation consultant. Physicians increasingly rely on functional evaluations and abrogate their responsibility to make the determination as to whether their patient is or is not ready to return to work. The ultimate decider has to be the patient's physician, who has the experience and training to judge. Deferring the decision to an objective "block box" test limits the flexibility of the system in the name of objectivity.

## William D. Sommerness:

The most important testimony is that from the physical therapist and occupational therapist, because they are the ones who have the "hands on" experience with the client and can do the best job of explaining the client's attitude and performance. With the exception of physiatrists, the least helpful testimonies are those from the physicians because they simply do not have the time or background with the client to adequately and competently testify as to the client's realistic abilities. Except for clear cases (eg, major limb damage, motor dysfunction) the utilization of a physician's opinion is almost meaningless. On the other hand, how can anyone competently and adequately analyze or realistically evaluate an employee's actual work status without the testimony of physical therapists and occupational therapists?

## Ronald Fischer:

In one crucial area of the workers' compensation decision-making process, the defense especially is at a distinct disadvantage. Many workers' compensation disputes arise out of hard fought issues of medical causation (eg, Did the employee's heart attack result from his physical exertion at work?) and the nature and the extent of disability (Is the employee able to work or is he malingering?). Judges generally consider the opinion of treating physicians to be persuasive over that of independent medical physicians hired by the employer. The employer may have hired the Chief of Orthopedic Surgery at the Mayo Clinic to independently examine an injured employee for back complaints, and yet the judge will be inclined to adopt the opinion of the treating local physician without regard to any disparity between the medical witnesses' experience or training, because the treating physician has seen the employee more often than the independent examiner and is therefore arguably in a better position to evaluate the employee.

Generally, medical physicians do not like to become involved as medical witnesses in personal injury lawsuits. Therefore, an employer is often relegated to call on the services of a certain number of physicians who have devoted a substantial portion of their practice to independent medical evaluations. However well qualified and highly principled, the independent examining physician may often appear to the judge as a "hired gun" when compared to the kindly treating physician who testifies for his patient.

Unfortunately, the treating physician is not an unbiased arbiter, but is generally sympathetically predisposed to his patient. The employee is his patient. His economic self-interest motivates the treating physician to satisfy his patients needs whether the need is to provide necessary medical treatment or to provide the supportive opinion necessary to a workers' compensation claim. Moreover, treating physicians are not trained to clinically "cross-examine" their patients. They will generally accept their patient's history and subjective complaints without critical evaluation. In that fashion, questionable subjective claims are legitimatized and corroborated by the physician.

The tactical disadvantage imposed on the defense is also procedural. Under the rules of practice in many jurisdictions an employer or his attorney cannot informally discuss an employee's disability with the treating physician during a litigation. The employee and his attorney are not similarly restrained. By the time of trial, the treating physician will therefore often have been influenced by conversations with his patient or the patient's attorney to accept the employee's position of the case. Conversely, the treating physician will often regard the employer's attorney as a hostile adversary whose purpose in appearing in the courtroom includes possibly embarrassing the physician or attacking the physician's testimony through cross-examination. In such circumstances, the treating physician will often hold fast in the courtroom to opinions and conclusions that initially, during the early stages of the physician's evaluation of the patient's condition, were only tentative and formed on information that was possibly mistaken or incomplete.

The particular issue involved in the identity of the employee's witnesses often dictates the defense counsel's choice of witness. For example, if the issue of the case is the necessity and reasonableness of further chiropractic care, the defense counsel will often call on the opinion and testimony of a licensed chiropractor. Physical therapists will be particularly helpful on issues of rehabilitation and in explaining physical limitations resulting from an injury. As a general rule of thumb, it is still perhaps true that physicians are considered more persuasive than other medical professionals when testifying. This assumes, of course, that the physician is testifying on an issue where he or she has the necessary background and training to qualify as an expert witness. Parenthetically, physicians are often less willing to testify than other medical care professionals, and when the physician does agree to testify, he or she is often not as willing to devote the time and effort that counsel may consider necessary in adequately preparing for the evaluation of a case and testimony. Therefore, if another medical care professional is available who is qualified to testify in a particular issue, I am very happy to use such testimony in lieu of a physician's testimony.

# Section 25

# The Main Feature: The Employee

*Margot Miller*

Of primary importance in the return-to-work process is the employee. If the employee does not understand his or her situation and cooperate, then none of the professionals involved in his or her rehabilitation will be able to effect a positive solution. The total rehabilitation plan must be the employee's plan as well or a positive outcome will not occur.

Too often in the return-to-work process the employee is not recognized as a full partner. The team approach is often mentioned and understood to include the therapist, the rehabilitation counselor, the physician, and the employer. Certainly, effective intervention requires a comprehensive team approach, but the most valuable team member—the employee—must not be forgotten.

What is the employee's perspective in the process of returning to the workplace? The first change is that from the "worker" image to that of a "patient" or "client." The employee is now part of a "system." Observations indicate that the worker-client falls into four models: able clients, unable clients, fearful clients, and unmotivated clients.

## ABLE CLIENTS

Able clients spend the least amount of time in the workers' compensation system with the resulting least amount of lost work time. Able clients have a strong desire to return to work. Even though they are not working immediately after injury, they consistently follow exercise programs and have less complications resulting from decreased flexibility, strength, and endurance. Able clients have the physical capabilities and the psychological strengths that allow return to work. Return to work is thus achieved before a delayed recovery occurs.[1-3]

Able clients present as mature, independent persons with strong self-concepts. These characteristics enable them to maintain control once injury occurs and to deal with temporary disability in a healthy manner. Able clients realize they are still able to function despite symptoms. In conversation with Dan, an able client, he puts it this way: "I knew if I didn't remain active I would lose strength and be worse off. Now I'm back working and it feels good."

**Case Example**

Carl is a 52-year-old man who works for a major newspaper in a small city. He has worked in maintenance for 22 years. He has an excellent work record and is a dependable employee.

Carl sustained a low back strain while lifting a 60-lb file cabinet at work one evening. He sought medical attention the next day. After 1 week of rest he had physical therapy on a daily basis as an outpatient. Treatment consisted of modalities as well as gentle trunk stretching techniques. As symptoms decreased and Carl became more flexible, strengthening exercises were added to his home program. Carl also attended a back rehabilitation program for further education in proper body mechanics and appropriate exercises.

Before return to work 4 weeks after the injury, Carl had a functional capacities assessment to determine his capabilities and limitations. A job analysis was provided by the employer. The evaluation indicated that Carl could perform safe lifting and carrying up to and including 50 lb. Carl stated that the only lifting he would have trouble with would be the occasional moving of heavy cabinets by himself. With help, this could be done as well. Carl had no difficulty in the areas of prolonged standing, repeated bending, walking, or overhead work required on the job. Carl successfully returned to work after injury and was able to perform all job tasks with the exception of lifting more than 50 lb. One month after return to work, lifting levels were re-evaluated. Carl could safely lift and carry 60 lb and thus perform his full job without restrictions.

Carl's successful return to work illustrates the able client's total commitment to returning to work. He was willing to follow a consistent program. He made continuous gains and was able to return to work in a matter of weeks.

## UNABLE CLIENTS

Unable clients clearly do not have the physical capabilities to return to work. This is usually the more complicated injury involving sophisticated testing procedures, lengthy treatment regimens, and, possibly, surgical intervention. A direct effect is greater lost work time and subsequent increased workers' compensation costs. For the client the delay means more time for weakness, tightness, and deconditioning to take place. The eventual outcome is that the worker lacks the physical requirements to return to work.

The employer may do two things at this point to allow unable clients to return to some form of work. The first involves modifying the job; that is, omitting those job duties that the client cannot physically perform. The client would participate in a functional capacities assessment to determine specific capabilities and limitations. The job analysis would be reviewed and the client would return to work performing only those duties he or she had the capabilities to perform.

The second involves modifying the worksite; that is, altering the worksite so the employee can still perform his or her regular job. Examples would be raising the height of a worktable to eliminate a forward trunk position or eliminating lifting from the floor by putting loads on a 12 in raised platform.

In either case, whether modifying the job or modifying the worksite, the employee is able to return to work.

Isernhagen[4] emphasizes every effort should be made to match an employee's physical abilities and restrictions to the actual work. If, without modification, the potential exists for reinjury, then individual job change must be made. In this way an unable client has the potential to become an able client. In conversation with Ed, a client whose employer modified his job so he could return to work, he states, "My employer was great. Modifying the job so I could return to work was the best thing for me. I know this isn't always possible and I feel very lucky to be working again."

### Case Example

Steve is a 31-year-old man who has worked in construction since he was 18 years old. He has been with his current employer for 6 years. He was injured falling from a scaffold. As a result of injury he had a lumbar fusion. Steve was told by his physician he would no longer be able to do heavy construction work. The employer has no light duty construction work. Steve has had no schooling other than a high school education. He enjoyed the physical exertion of construction work as well as working outdoors. He has been off work for 8 months and must now look into other job opportunities. This becomes a complex problem for both Steve and his family.

Steve's case illustrates the difficulties encountered when a client lacks the physical requirements of the job. Modifying the job or altering the worksite is not always possible. New careers may need to be pursued in areas oftentimes avoided previously, as in desk work for Steve. Feelings of anger and depression take hold. Both client and family are influenced by the required changes and a complex picture evolves. In Steve's case both he and his wife sought counseling to help them through the injury process. Understanding their feelings of anger and sorrow was the first step. Acceptance would come as would a new career for Steve in drafting. Heavy physical work was not possible, but in drafting Steve could use his past expertise and experience. Steve was once again a productive, capable worker.

## FEARFUL CLIENTS

Fearful clients represent a large group of injured workers. They are categorized as fearful because they have capabilities mixed with fear, uncertainty, and anxiety about returning to work.[1,2,5] Concerns of fearful clients might be: What if I reinjure myself? How much can I lift? Will my co-workers give me a bad time if I can't keep up?

In conversation with Jerry, a typical fearful client, he states he worries about finding another job and wonders who would hire him. "The stress is tremendous. I worry and only get more frustrated. I wonder if and when I will be working again?"

Successful return to work for fearful clients involves a team effort. The physician, the employer, the rehabilitation counselor, and therapists all need to work together to coordinate a return-to-work effort. The physical symptoms as well as the psychological perspective of the injured worker need to be addressed.

The most positive approach for fearful clients is to have them participate in a work hardening program after the necessary acute therapy. (See Section 15, *supra*.) Such a program would harden the client and enable him or her to return to work. In work hardening, the client may participate in a psychological component and address the fear and uncertainty he or she has. A functional capacities assessment before the the work hardening program would outline specific strengths and weaknesses, and define an effective plan to increase work tolerance. In a short period of time, usually 4 to 6 weeks, the fearful client will be participating in a program to increase strength, flexibility, and endurance as well as to increase knowledge, motivation, and coping mechanisms. The fearful client gradually regains confidence in his or her abilities and is prepared for return to work. By combining effective treatment for both the physical and psychological symptoms, the client will make the necessary gains to enable return.

The fearful client can become the able or the unable client, depending on the total rehabilitation and its institution. The key to a successful return is a team effort with all parties working toward an early and safe return to the workplace for the employee.

**Case Example**

Linda is a 43-year-old woman who works as a nursing assistant in a large hospital. She has worked for her current employer for 12 years. She is an excellent employee, and the employer has indicated he will do what he can to get Linda back to work. Linda sustained a neck and upper back strain while transferring a patient. After 2 weeks of rest, Linda was sent to physical therapy for treatment She has made small gains but has difficulty following the exercises because of discomfort. She is very fearful of pulling something and hurting herself more. Her physician says if it hurts, don't do it. She spends most of her day lying down. Linda is not only losing in strength, flexibility, and general endurance but in her feelings of self-confidence as well. She becomes less and less a capable and functioning person.

At the recommendation of her rehabilitation counselor, Linda participated in a functional capacities assessment to outline her capabilities and limitations. Because of significant physical limitations, a work hardening program was recommended. Linda's fear of increasing discomfort and hurting herself keeps her from performing to her fullest. Because of inactivity she is physically deconditioned. A job analysis is obtained from Linda's employer. In work hardening she participates in exercises to increase her flexibility, strength, and general condition as well as job simulation activities, simulating actual job duties of a nursing assistant. She is instructed in proper lifting and transfer techniques. She receives positive feedback on her daily gains. Her fear and anxiety are further addressed by a psychologist. She is instructed in relaxation techniques. She begins to understand the process that has been going on since injury. Linda gradually becomes a more physically capable person and gains in self-esteem and self-confidence as well. She is able to return to work part time and work up to full time eventually performing full job duties.

Linda's case illustrates the transition from fearful client to able client with a team approach and a successful work hardening program. Gradually

Linda sees an increase in her physical abilities. Her fear and uncertainty are addressed. She is physically and emotionally prepared to return to work.

## UNMOTIVATED CLIENTS

The last and smallest group of injured workers are unmotivated clients. They are categorized as unmotivated as they do not want to return to work for a number of underlying reasons. Ron typifies this client and he expresses his concerns in conversation: "If I return to work I will not be doing a job I want to do but will be forced to do a job I can do." What Ron doesn't realize is he is unconsciously preventing a successful return to work for himself.

Unmotivated workers seek pain or injury as a solution to problems. They may not like their current job or may be having marital or other relationship problems. The injury takes these clients away from the job they didn't like anyway or allows them to be less responsible family members, for example. Unmotivated clients become less social and withdraw. They become accustomed to the invalid role. They are not motivated to return to work or to increase activity level. During a functional evaluation, inconsistencies are often seen. Unmotivated clients will not see capabilities in a positive light and will often concentrate on their limitations. It is essential to recognize and identify the possible underlying reasons for such behavior if gains are to be made with the unmotivated client.

The unmotivated client is not necessarily consciously manipulating; he or she is not aware that the pain keeps him or her from dealing with a job that he or she does not like anyway. A malingerer, however, consciously uses pain and symptoms; he or she deliberately uses the injury for gain. Killian[6] (see also Sections 17 and 18) reports the key in identifying a true malingerer is the conscious and intentional simulation of symptoms or disability. The number of true malingerers is very small, and it must be realized that for most unmotivated workers, the process is unconscious. The psychological needs of this injured worker must be addressed early after injury to effect a positive solution.

### Case Example

Tom is a 50-year-old man who works as an inspector for a grain elevator company in a large port city. He has been with the company for 22 years and is a dependable worker. Because of a decline in production, Tom has seen several of his co-workers lacking in seniority laid off. He has doubts whether he will be allowed to continue working as there is talk about another layoff.

His wife works part time earning minimum wage as a waitress. Tom has five children; two are in college. He is already anticipating the financial strain if he were not working.

Tom sustained a thoracic strain after falling from a stepladder. He sought medical attention and is to be on bed rest for 2 weeks. He reports symptoms of excruciating pain in the back radiating into the lateral and anterior chest. He does not benefit from a trial of outpatient physical therapy as he complained the symptoms persisted. Further x-ray films and testing procedures were ordered. There were no physical findings necessitating further medical procedures. Tom is feeling the financial strain of not working; however, he is receiving workers' compensation, which is more advantageous to him than unemployment. There is much family tension and Tom reports marital difficulties to his physician. Nine months after injury for a thoracic strain, Tom is still not working and reports the excruciating pain persists. The physician and physical therapist have no clear answer regarding what should come next.

This is clearly a case where Tom is unconsciously using the injury and subsequent symptoms to avoid returning to work and looking at a possible layoff. Unless the psychological needs are met for this injured worker, a decrease in symptoms and increase in activity level are nearly impossible. The injury allows Tom to continue in his role as a patient, to remain on workers' compensation, and to avoid a possible layoff.

Litigation may be inevitable in such cases with disagreement most likely between consulting physicians regarding the percentage of disability. Once a settlement is made and case resolution occurs, the client again faces the necessity of work. After case settlement Tom became the manager of a small restaurant. He took a night course in business management and for the first time in months actually looked forward to work. Psychological intervention helped him understand the injury process. A work hardening program helped him increase his physical abilities. Tom eventually became a productive and capable person.

## SUMMARY

All professionals must remember the importance of the employee in the return-to-work process. In the team approach, the employee stands out as the most valuable player. Without employee understanding and cooperation, successful return to work will be jeopardized. All professionals can work together with the employee to effect a positive solution.

## REFERENCES

1. Hirschfeld AH, Behan RC: The accident process. I. Etiological considerations of industrial injuries. *JAMA* 1963;186:193–199.
2. Behan RC, Hirschfeld AH: The accident process. II. Toward more rational treatment of industrial injuries. *JAMA* 1963;186:300–306.
3. Derebery VJ, Tullis WH: Delayed recovery in a patient with a work compensable injury. *J Occup Med* 1983;25:11.
4. Isernhagen S: The physical therapist and injured worker: The right match. *Whirlpool*, Winter 1985.
5. Van Wagner R, Racer H: Work-related injury: A community context. *Network* 1984;1:1.
6. Killian L: Delayed recovery: Is it psychological? *Network* 1985;2:3.

# PART X

# Future Challenges in Occupational Medicine[*]

*Susan J. Isernhagen*

## AN EMERGING FORCE

The future of any specialty is dependent on the commitment of those professionals delivering service and creating standards. The rapid growth of occupational medicine will continue for many years and eventually result in a large body of knowledge and research. Educational criteria for professionals interested in this specialty will be developed and implemented.

In the management of physical work injury there are three primary areas that will affect future growth and development: commitment of medical professionals, external forces and regulations, and changing industrial factors.

## COMMITMENT OF PROFESSIONALS

In this book, the concept that a work injury is, in fact, a medical problem has been emphasized. Because it is a physical injury, medical practitioners will be primarily responsible for care and treatment of the person suffering such injury. Whereas medical care has always been dedicated to improving the health of patients, occupational medical care also is dedicated to improving the worker's health and the worker's physical relationship to work. Therefore, the internal commitment of both medicine and industry to work together and individually for safe worksites and quality care of the injured worker will influence the future of occupational medicine.

Each section in this book has focused on current practices and future challenges. In summarizing, the authors have all stated their commitment

---

[*]The author gratefully acknowledges the contributions of Suzanne H. Rodgers and Naomi Elizabeth McCabe to this section.

to progress, early and thorough care of the worker, prevention of injury through education and ergonomics, recognition of the complexity of work injury, and dedication to the team approach.

These mutual goals are the cornerstone of effective physical work injury management. Both returning people to work and assessing people's suitability for a specific job may involve therapists, physicians, nurses, medical specialists, industrial specialists, industrial supervisors, engineers, and ergonomists. Although not all of these people may be available in a given workplace, each discipline has valuable information.

Quality assurance programs will be important in the future of medical-industrial systems. Both medicine and industry must be satisfied that the highest quality of preventive or injury care is given. By studying outcomes, quality of service, and individual aspects of care delivery, effective methods and programs can be defined. It will be extremely important that guidelines are actually set by medicine and industry to avoid negative restrictions set by outside agencies.

Qualifications of personnel involved in occupational medicine will become easier to define in the coming years. Generalists can become specialists in either medicine or industry, but specific training and experience are necessary to ensure the highest quality services. Formal educational programs will be developed to train medical personnel in work injury management. Both medical and industrial programs in occupational health and medicine will incorporate information on each other's disciplines. Occupational medicine will have industrial components for medical personnel and medical components for industrial personnel.

Technology also is advancing. Computers, video recorders, and measurement instrumentation in functional activity will aid progressive specialists in implementing the advanced programs now being developed. Communication will be facilitated by computerization to provide clear and immediate exchange of information among professionals in work injury management.

## EXTERNAL FORCES AND REGULATIONS

### Reimbursement

Workers' compensation insurance companies are quickly reaching conclusions regarding the necessity for curbing reimbursement. These companies will decide either to set criteria for work injury programs or to opt for a mandated ceiling on reimbursement for such services.

Inexpensive and ineffective programs could flourish under arbitrary reimbursement limits. The ceiling could prevent the high quality occupational medicine programs from advancing. Therefore, a challenge for the future will be the cooperation of medicine and industry in assisting insurance companies to develop criteria for evaluation of effective, safe, and productive occupational medicine programs. Long-term results should outweigh short-term costs.

As for insurance companies, economic problems are one of the main concerns for state workers' compensation programs. In states where rehabilitation consultants are prevalent, their cost effectiveness is being studied. The rapid growth of medical occupational specialties, especially in rehabilitation (functional assessments and work hardening), has also caused concern and demand for outcome criteria. Personnel involved in workers' compensation programs, being involved in a legislative entity, will need to seek input from, and also educate, medical providers, industry, and unions.

**Safety**

Return to work with worker safety rather than payment of disability claims should be the prime concern for workers' compensation systems. If disabilities are identified, settlement should also include a way to facilitate a more productive life rather than an end to a career.

The US government is involved in research on accident rates and prevention techniques. The future should see more practical guidelines being set for industry. The challenge for government will be to provide industrial evaluation and enforcement mechanisms without becoming restrictive in innovations in occupational medicine development.

Industry's challenge will be to encourage commitment to matching the work to the worker rather than the current philosophy of matching the worker to the work. Ergonomics should aid in reducing physical costs of heavy or repetitive work.

**Standards**

Medicine will continue to develop standardized diagnostic and treatment procedures for early, thorough physical restoration. Education of the worker in health practices should not only ensure freedom from specific work injuries but also improve the overall health of workers and their families.

Standards and uniformity will be important in the development of occupational medicine programs. Across the United States, various professions and vested parties need to pursue standards without giving up their individual ways of offering services. The large variations in current service provisions will gradually change to fewer but stronger options in delivery of service.

## Management

External forces will include cost control as well as medical excellence thereby facilitating a total business approach. Industrial medicine and safety can be combined with effective business practices without losing the quality edge. Survival of the best occupational medicine programs will depend not only on their provision of service but also on their ability to maintain business viability.

## CHANGING INDUSTRIAL FACTORS

A changing work force will color effective management of physical work injury. For the next 10 to 15 years, there will be an older work force in most large industries for the following reasons:

1. A reduced number of young people will be entering the work force because of reduced population growth during the sixties and seventies.
2. There will be an increased life span and possibly a later retirement age for older workers.
3. Growth in manufacturing industries will be reduced, so there will be less employment of younger workers and less turnover of older ones.
4. There will be a change in the intensity of manual work with lighter, repetitive work becoming more prevalent. This is an outcome of changing from an industrial to an information society. This will allow older workers to remain in the work force to a greater age. It also will change the type of work injuries from a heavy emphasis on manual and lifting injuries to greater repetitive trauma syndromes.

The older work force will be subject to the types of job demands that may have been handled by younger workers in the past. In either repetition or weight capacities this may lead to a mismatch of worker and work. A coordinated approach to identify tasks that aggravate these symptoms and to design them out of new jobs and reduce them in existing jobs will be the most effective approach for keeping skilled workers on the job and free of injuries.

The trend toward a higher percentage of deconditioned population will match the new emphasis on sedentary jobs. However, lesser physical strength will not be matched with lesser physical output and the percentage of risk of injury may be approximately the same.

Therefore, the structure of the work and the worker will need to be evaluated throughout the years for changed or improved methods of work injury management.

## SUMMARY

In summary, the rapidly expanding occupational medicine field has made a positive impact in reducing work injury and preventing its chronic long-term effects. The team approach of thorough medical care with industrial communication is allowing mutual goals to be reached.

The availability of information of successful programs throughout the world is a factor in improving work injury management in all countries. No longer do local, regional, or state, or even national problems have to be solved locally. The trend is for communication and cooperation for the attainable goal of effective work injury management programs.

# Index

## A

Able clients, 347–348
Absenteeism, health awareness programs and, 31
Action limit, 86–87
Acupuncture, as cure for smoking, 32
Acute injury, 304, 305
Aerobic capacity assessment, 74
Aerobic exercise, 22
  as phase in fitness program, 31–32
Age. *See also* Elderly workers
  as factor in ability to move weight, climb stairs, and move on balance beam, 184–191
Aging process, 59
Airline industry, 20
Alcoholic workers, 249
American College of Sports Medicine weight control guidelines, 33
American Occupational Medicine Association, 299
Anaerobic metabolism, 71
Anatomy instruction, 15
Anderson, Charles K., 92, 100
Anthropometric information
  musculoskeletal injury risk and, 47
  value in lessening cumulative trauma of, 63–64
Anxiety regarding reinjury, 255
Apts, David, 19, 22, 26
Asbestos worker prescreening, 108
Assembly boards work simulation, *220*
Attorney
  role in work injury cases, 338, 342–344
  workers' compensation and role of, 6
Attorney's perspective
  on others in workers' compensation system, 344–346
  on role regarding clients, 341–342
  on workers' compensation and legal process, 339–340
Australia, 56

## B

Back Depression Inventory, 291
Back injury. *See also* Low back pain
  frequency and cost of, 39–40
  low, 39, 59
  prevention and workstation design, 85
  return to work after, 134
  use of X-rays to assess risk of, 97
  worker training to reduce, 45–48

NOTE: Page numbers in Italics refer to artwork in the text.

Back pain. *See also* Low back pain
  causes of low, 42–43
  pre-employment screening for, 95, 96
Back schools. *See also* Health education programs; Injury prevention programs
  ergonomic intervention as part of program in, 85
  future outlook for, 28
  injury prevention effectiveness of, 19–22, 134
  teachers for, 23, 199
  topics to cover in, 23–25
Balance
  age and gender as factor in, 184, 188–190
  testing, 148, 186
Base line functional analysis, 208–210
Behavior management, in functional capacity evaluation, 161
Behavioral assessment, 252
Biodex, 236
Biofeedback techniques, 289
Biomechanical fatigue, 82
Biomechanical stress, 41
  design principles and, 45
  sources of, 44
Bly, J. L., 31
Bona fide occupational qualifications (BFOQs), 110
Brisson, Jean, 120
Broderick, Tom, 125
Bulging disk syndrome, 59
Bullock, Margaret I., 9

## C

Cardiac risk factors, 30–31
Cardiac stress test, 31
Cardiopulmonary evaluation
  as part of functional capacity evaluation, 165–166
  for susceptibility to cumulative trauma, 62
Cardiovascular disease risk factors, 32

Cardiovascular fitness
  in chronic pain patients, 288
  instruction for, 22
Carpal tunnel syndrome, 43, 58. *See also* Repetition strain injury.
  ergonomic intervention regarding, 76–78
Casper, Jack, 175
Chaffin, Donald B., 86, 93, 98, 100
Chelsea Back Program, 21
Chronic injury, 304, 305
Chronic pain
  attributes, 284
  history of treatment models for, 284–285
  program planning for, 286–287
  work injury and, 283–284
Chronic pain management
  case study of, 292–294
  coordination of services for, 290
  education of patient regarding, 287–288
  at Indiana Center for Rehabilitation Medicine, 290–292
  physical intervention for, 288–289
  psychological intervention for, 289–290
Circulation, evaluation of cardiopulmonary, 63
Circulatory fatigue, 82
Clinics, 119
Cognitive assessment, 252
Communication
  employer–medical provider, 323–325
  of musculoskeletal safety concepts, 48–49, 56
  by therapists, 332
Comparative functional analysis, 223
Comprehensive psychophysical model for work hardening
  case closure, 225
  daily work schedules in, 212–215
  environmental problems of, 224–227
  identification of work–related rehabilitation goals in, 210–211
  initial evaluation, 206–210

organizational problems of, 224
overview of, 204–206
professional application of, 223–224
program philosophies of, 211–223
Computer technology, used for information dissemination, 49–50
Conferences, 336–337
Connors, Bart, 335
Conservative care, for spinal injury, 237
Construction electrician work simulation, *217*
Construction framing carpenter work simulation, *219*
Continuous effort duration, 69–71
Conversion disorder, 265, 267
Conversion reactions, 248
Cool-down phase of exercise program, 32
Coordination testing, 147–148, *159*
Courtney, James, III, 339, 341–344
Courtroom testimony, regarding functional capacity evaluation, 166–167
Cumulative trauma
anthropometric data and, 63–64
causes of, 56–58
decreasing potential for, 60–61
definition of, 55
individual susceptibility to, 61–63
types of, 58–59

## D

Deconditioning syndrome, 234
deCuervain's syndrome, 58
Denmark, 55
Dependent individuals, disability in, 248
Deposition regarding functional capacity evaluation, 167
Depression, 247-248
Design, workplace, 13, 15. *See also* Workstation design
Diagnosis, 299–300
Diagnostic decision tree, 262
Diagnostic workup, 300–301
Directed occupational medicine examination (DOME), 107–109, 111

Disability determinations, 140
job classifications and, 177–180
overview of, 175–176
Social Security, 176–177
Disabled workers
employer prejudice regarding, 225
health education in workplace for, 11
Discrimination, employment, 106, 110
Dissemination limitations, regarding musculoskeletal problems, 49–50
Drug screening, 108
Duluth Clinic, 120–123
Duration of continuous effort, 69–71
Dynamic tests, 99–100
Dynamometer, 73

## E

Educational programs. *See* Health education programs.
Educational Resource Centers (ERCs), 45, 46
Effort intensity, 66–69
Elderly workers
balancing capabilities of, 188–191
disability in, 248–249
health education in workplace for, 10–11
industrial factors and, 358–359
risk level for, 184
stair climbing ability of, 186–188
weight capacities of, 186, 187
Emotional stress, 13
Employee
as able client, 347–348
ergonomic principles and, 54
as fearful client, 350–352
as unable client, 349–350
as unmotivated client, 352–353
workers' compensation system and, 4, 339
Employee attitude regarding health education, 18
Employee motivation regarding injury prevention programs, 26
Employee screening, 47–48. *See also* Pre-employment screening

Employee strength assessment, 72–74
Employer perception of workers' compensation system, 339–340
Employer work–injury management. *See also* Work–injury management
  and communication with medical providers, 323–325
  importance of good communication by, 328
  injury prevention and, 317–318
  injury report and, 318–319
  insurers and, 328–329
  job modification and, 325
  labor unions and 327–328
  lost time and, 319
  peer pressure and, 325–327
  and restricted duty program, 322–323
  worker support and, 319, 322
Employment discrimination, 106, 110
Employment selection procedures. *See* Pre-employment screening
Endurance testing
  in functional capacity evaluation, 146–147
  for work capability assessment, 73
Epicondylitis, 58–59. *See also* Repetition strain injury
Equal Employment Opportunity Commission (EEOC) guidelines, 106
ERCs (Educational Resource Centers), 45–46
Ergonomic prevention programs, 15
Ergonomics
  availability of information regarding, 49
  cumulative trauma and, 54–64
  definition of, 54
  international studies in, 55–56
  interventions in workplace, 74–78
  used in job tool redesign, 44–45
  used in workplace redesign, 47–48
  value of, 9–11, 16
Evaluation, psychological, 252
Exercise breaks, 16. *See also* Pause gymnastics
Exercise progams. *See* Fitness programs

Exertion, cumulative trauma and, 57
Expectational horizon, 276
Expert witnessess, 166–167
Extremity disorders, 43
Extremity neurological function, 234

# F

Factitious disorder
  chronic, with physical symptoms, 269
  definition of, 268–269
  malingering vs., 267
  with psychological symptoms, 269
  types of, 269
Fantel, J.D., 97
Fatigue
  as warning sign for repetitive strain injury, 82-83
  worker positioning to decrease, 84
  workstation design to decrease, 84
Fearful clients, 350–352
Feasibility for employment, 197
Female workers
  balancing capabilities of, 188–190
  risk level for, 184
  stair climbing ability for, 186–188
  weight capacities for, 186, 187
Fibrositis, 59
Finland, 56
First aid report form, 318–319, *320-321*
Fischer, Ronald, 339–340, 342, 343, 345
Fitness
  benefits of, 30–31
  education regarding, 13
  evaluation of, 31
  low back pain and, 95
Fitness programs
  components of, 31–32
  effectiveness in injury prevention, 22–23
  table of, *34*
Fleeson, William, 106, 298
Forward head position, 59
Frymoyer, J. W., 95

Functional Capacities Assessment (FCA)
  age and gender correlations found using, 184–191
  case studies illustrating cost effectiveness of, 181–183
  overview of, 181, 185
Functional capacity evaluation (FCE), 137, 139-140
  adjuncts to, 164–166
  case study of, 169–174
  of chronic pain patients, 290
  definition of, 139
  deposition regarding, 167
  design of, 144-148
  evaluator perspective in, 149, 159–162
  for fearful clients, 350
  legal implications related to, 166–167
  limited, 143–144
  parameters of, 140–144
  patient data for, 150–151
  report formats for, 162–163
  summary report, 151–152
  test, 155–160
  therapist and 335–336
Functional capacity evaluator
  as expert witness, 166–167
  perspective of, 149, 159–162
  skills required for, 141–142
Functional restoration for chronic disorders, 237–241
Functional strength tests, 98, 145–146
  dynamic, 99–100
  isokinetic, 146
  isometric, 98–99, 146
  psychophysical, 100
Functionally fit for work, 206

## G

Ganglion cysts, 58
Gender
  as factor in ability to move weight, climb stairs, and move on balance beam, 184–191

General Mills, Inc. 30, 31
Gloves
  grip strength and, 68
  work affected by use of, 70
Goaling, 275–277, 280
Goals, 276
Gonodal irradiation, 97
Gore, Amanda, 35
Governmental agencies, workers' compensation system and, 5–6
Graded mastery techniques, 274, 277
Gravity, cumulative trauma and effect of, 58
Grip span, 66, 67, 70
Grip strength, 66–68, 70

## H

Hand conditions, 58–59
Hand coordination testing, 147, 159
Handgrips, 60
Handling. See Manual handling and lifting
Hari-Kari, 277
Hazard surveillance, to lesson musculoskeletal injury, 41
Health club membership, 23
Health education programs, 9–10
  content of, 15–16, 23–26
  effectiveness of, 22–23
  effects of legislation on, 16
  for elderly workers, 10–11
  employee attitude regarding, 18
  future challenges, 26–28
  in industry, 30–33
  limitations of, 17–18
  objectives of, 13–15
  planning for, 10–11
  for specific types of work, 11–13
  table of, 34
Health insurance costs, 27
Health risk appraisal, 31
Health surveillance, to lessen musculoskeletal injury, 40
Heavy work classification, 178
Helplessness, 260–261
High-risk jobs, 100, 102

Hospitals, 119
Hypertension management, 32–33
Hypnosis, as cure for smoking, 32
Hypochondriasis, 265
Hysterical neurosis—conversion type, 265
Hysterical personalities, 248

## I

Iatrogenic illness, 197
Imbrogno, Dean, 1, 2
Immature individuals, disability in, 248
Immobilization, 130–131
Indiana Center for Rehabilitation Medicine, 290–292
Individual characteristics, as pre-employment screening approach, 95
Industrial health programs. *See* Fitness programs; Health education programs
Industrial Medicine Center (Duluth, Minnesota), 106–107
Industrial on-site medical teams for work-injury treatment, 117
Information retention limits, 17–18
Injury prevention programs. *See also* Health education programs.
employee motivation regarding, 26
ergonomic intervention as part of, 85
formal, 317–318
Injury treatment. *See* Work-injury treatment
Insurance carriers
reimbursement by, 356–357
role in return to work of, 328–329
Intermittent work, 72
Irradiation, disease risk caused by, 97
Isernhagen, Susan J., 1, 19, 30, 35, 54, 80, 115, 127, 139, 184, 338, 349, 355
Isodynamic tests, 99
Isokinetic tests, 99
for functional capacity evaluation, 164–165
of trunk muscles, 100–101, 236
Isometric exercise, 131

Isometric testing, 98–99
for functional capacity evaluation, 165
of lifting, 234

## J

Jensen, Roger C., 39
Job analysis
importance of, 191
pre-employment screening and, 102
Job classifications, 177–180
Job demands, 65–66
duration of continuous effort and, 69–71
effort intensity as factor of, 66–69
work pattern as, 72
Job modifications, 140, 279, 325
Job redesign, 44–45, 65
Job restrictions, 333
Job rotation
employee interest in, 17
as means of copying with stress, 15–16
Job simulation
creativity in, 331
testing, 108–109
used in work hardening program, 216–223
Johnson, Laurie, 184
Johnson, Timothy R., 30
Johnson and Johnson, 31
Joints, evaluation of, 62–63

## K

Kelsey, J. L., 181
Keyboard operation stress, 81, 84. *See also* Repetition strain injury
Killian, Lynne E., 247, 352
Kin–Com, 236
Kishino, N., 99
Knee injury, 43

## L

Labor Department
guidelines on validation procedures, 103
job classifications by, 177–178

Labor unions
  light duty and requirements of, 310
  role in return to work by, 327–328
Large birdcage work simulation, 222
Lawyer. *See* Attorney
Learned helplessness, 259
Leg muscle strength improvement, 46
Legal action
  related disability determinations, 175–180
  related to functional capacity evaluations, 166–167
Legislation, educational programs and awareness of, 16
Lett, Carol Franz, 195
Leukemia, X-ray exposure during pregnancy and risk of childhood, 97
Lido, 236
Lifestyle and fitness, 30–31
Lifting. *See also* Manual handling and lifting
  age and gender and capacity for, 186, 187
  guidelines for, 86–87
  situations requiring, 87
Light duty, 309–310, 319
Light work classification, 178
Limited functional testing, 143–144. *See also* Functional capacity evaluation
Lloyd, D. C., 96
Low back pain, 39. *See also* Back pain.
  causes of, 42–43
  ergonomic interventions regarding, 74–76
  fitness and, 95
  pre-employment screening for likelihood of developing, 95, 96
Low sensitivity, 97
Low specificity, 97
Lumbar spine, 239
Lung disease, occupational, 39
Luopajarvi, Tuulikki, 55, 56

# M

McCabe, Naomi Elizabeth, 195, 355$n$
McIntyre, D. R., 99
Male workers
  balancing capabilities of 188–190
  stair climbing ability for, 186–188
  weight capacities for, 186, 187
Malingerer, 251, 266
Malingering, 4
  definition of, 266
  diagnosing, 267
  management of, 161, 251
  overview of, 251, 255
  secondary gain neurosis vs., 250
  symptom magnification syndrome vs., 267–268
Managers, health education programs for, 14–15
Mantoux test, 108
Manual handling and lifting, health education programs covering, 11, 16. *See also* Lifting
Matheson, Leonard N., 196, 257
Mattmiller, Bill, 20, 21
Maximal permissible limit (MPL), 86
Maximum medical improvement (MMI), 308
Mayer, Tom G., 232
Medical history and pre-employment screening, 96
Medical professionals, workers' compensation system and, 5
Medical team approach, 117
Medium work classification, 178
Miller, Margot, 181, 184
Minnesota Multiphasic Personality Inventory (MMPI), 291
Mistal, Mary A., 20, 22, 23, 27
Mokros, Karen, 184
Morris, Alan W., 80, 87, 92
Motion analysis, 165
Motivation, employee, regarding injury prevention programs, 26
Muscle evaluation, 62
Muscle strength measurements, 72–74
Muscle strength tests, 98, 100–101

Musculoskeletal injury. *See also* NIOSH strategy for reducing musculoskeletal injury, Spinal disorders, chronic.
  immediate care for, 127–128
  NIOSH strategy for reducing, 40–50
  prevention of, 22
Musculoskeletal system
  education in, 15
  evaluation of, 207, 208
Mynerts, Gunilla, 80, 85
Myofascial syndrome, 304

**N**

Nachemson, A., 197
Narrative reports, 163
National Institute for Occupational Safety and Health (NIOSH), 39–40, 93
National Safety Council Conference, 20
Neck problems, 59
Negative cognitive set, 261
Negative transposition, 273
Nelson, Roger M., 39
Nester, David E., 39
Neurophysiological fatigue, 83
Neurotic anxiety, 250
NIOSH (National Institute for Occupational Safety and Health), 39–40, 93
  Minerva project, 49
NIOSH strategy for reducing musculoskeletal injury
  by controlling occupational risk factors, 43–48
  by evaluating cause and effects, 42–43
  by increasing awareness and stimulating interventions, 48–50
  by refining surveillance systems, 40–42
*NIOSH Work Practices Guide for Manual Lifting*, 86, 87, 93, 99
Norway, 55

**O**

Objectives, 276
Occupational health, 1-2

Occupational health care, in Scandinavian countries, 55–56
Occupational lung disease, 39
Occupational medicine
  definition of, 2
  future challenges in, 355–359
  need for medical school courses in, 49
Occupational medicine examination, directed, 107–109, 111
Occupational medicine health systems, 119
Occupational medicine hospitals, 119
Occupational medicine physician, 118–119
  ending care by, 307–308
  importance of using, 298
  locating, 299
  ongoing care by, 304–307
  transmitting information to employer by, 303–304
  treatment plan of, 301–302
Occupational overuse syndrome. *See* cumulative trauma
Occupational risk factor control, 43–44
  by ergonomic job/tool redesign, 44–45
  by selection and placement procedures, 47–48
  by worker training for good work practices, 45–47
Occupational Safety and Health Administration (OSHA), 46
Occupational therapist, 5
  and role in return to work, 331–337
Occupational therapy, 197
  for purpose of work hardening, 200–202
*OFCCP v. EE Black, Ltd.*, 97
On-site medical teams for work-injury treatment, 117
On-site physical therapy, 128
Overexertion, as most common source of compensable injury, 92

## P

Pace, 147
  age and, 187, 188
Pacing, stress related to, 12
Pain, 161. *See also* Chronic pain; Chronic pain management
Pause gymnastics programs
  benefits of, 37–38
  methods for introducing, 35–37
Peer pressure, 325–327
Percentage of maximum voluntary contraction strength, 68–69
Perception Validation Interview, 278
Person, Peter, 120, 123
Physical examination
  importance in pre-employment of, 96
  preplacement, 106–111
Physical fitness. *See* Fitness
Physical fitness classes, 16
Physical stress
  in sedentary work, 12–13
  teaching methods of coping with, 15–16
Physical therapists, 5
  as instructor in injury prevention programs, 23, 27, 28
  and role in immediate care, 127–133
  role in return to work of, 331–337
Physical therapy
  contract services for, 129
  interventions in immediate care, 130–132
  on-site, 128
  proximal, 129
  for purpose of work hardening, 198–200
  trends in, 132–133
Physical work-injury management, 2. *See also* Work-injury treatment
Pinch grips, 66
Pizatella, Timothy J., 39
Polinsky Medical Rehabilitation Center, 184, 185
Porter, Ronald W., 21–23, 27
Posterior disk bulges, 59

Posture
  instruction in, 15
  neck and shoulder problems from bad, 59
  stressful, 57
Power grip strength, 66–69
Pre-employment screening, 91–93
  alternative forms of, 95–101
  development and implementation of program for, 101–103
  selection procedure evaluation for, 93–95
  with X-rays, 97, 108
Pregnancy and X-rays, 97
Preplacement physical examinations, 106–107
  directed occupational medicine examination as, 107–109, 111
  problems with standard, 107
  reasons for, 109–111
PRIDE program, 235–236, 238, 239, 241
Productive Rehabilitation Institute of Dallas for Ergonomics (PRIDE). *See* PRIDE program
*Proposed National Strategy for the Prevention of Musculoskeletal Injuries,* 40
Prospective study, 103
Pseudo-self-sufficient individuals, 248
Psychogenic pain disorder, 265
Psychological barriers to recovery
  post-traumatic factors, 250–251
  pre-existing factors, 247–249
  traumatic factors, 249–250
Psychological stress
  in repetitive work, 12
  in sedentary work, 13
  teaching methods of coping with, 15–16
Psychologist
  consultation by, 252–253
  evaluation by, 252
  treatment by, 254–256
Psychophysical tests, 100

Psychosocial problems as barrier to recovery, 236, 239
Purpose, 276

## R

Rabadi, R., 68
Radiologic measurements, 47
Railway workers' training in materials handling, 45
Ramp and bins work simulation, 221
Recovery time between exertions, 71
Rehabilitation Act of 1973, 97, 101
Rehabilitation consultant, 5, 311
  controlling function of, 315–316
  leading function of, 314–315
  organizational function of, 313–314
  planning function of, 311–313
  workers' compensation system and, 5
Rehabilitation goals, 210–211, 305–307
Relaxation techniques
  for chronic pain management, 289
  education in, 13, 15–16
Repetition strain injury. See also Cumulative trauma.
  contributing factors to, 81–82
  definition of, 81
  fatigue as warning sign of, 82–83
  workplace modification for decrease in likelihood of, 83–84
Repetitive work, 12
Report formats for functional capacity evaluation, 163, 164
Reproducibility of measurement, 235
Residual functional capacity (RFC), importance of assessment of, 175–176, 180
Response-outcome independence, 260
Restricted duty, 309–310, 319
  program for, 322–323
Retrospective validation study, 103
Return to work, 137
  assessment of readiness for, 137, 140. See also Functional capacity evaluation.
  attorney's role in, 342–344
  early, 309–310
  employer's role in, 317–330
  philosophy regarding, 116, 125, 131–132
  physician's role in, 298–310
  rate, 241
  recommendations, 191
  resistance to, 307–308
  therapist's role in, 334–337
  work capacity services and, 226–227
Rockey, P. H., 97
Rodgers, Suzanne H., 65, 355n
Rounded shoulder posture, 59
Rowe, Miriam, 80, 95
Rust International Corporation, 125

## S

Sabotage Scale, 291
Safe body mechanics, 148
Safety offices, 11
Safety programs, 317–318, 357
Safeway, 21
Samuri game, 277
Saunders, Duane, 22, 24, 27
Sawhill, J. A., 99
Scandinavian countries, occupational health care in, 55–56
Seating requirements, 16
Secondary gain neurosis, 250, 251, 255
Sedentary work
  classification, 178
  health education programs covering, 12–13
Selby, Nancy C., 21, 22, 27
Self-determination, 162
Semiskilled jobs, 178
Sheldahl, L. M., 198
Shoulder problems, 59
Sickness Impact Profile, 291
Sit/stand jobs, 179
Sitting position, 16
Skill categories for jobs, 178
Skilled jobs, 178
Slosson Intelligence Test, 203
Smith, Patricia, 278
Smoking, 30–32

Smoking cessation, 32
Snook, Stover, 86, 95, 100
Social assessment, 252
Social Security, 176–177
Social Security Administration, 176
  job classification by, 177–178
Social worker, as member of chronic pain management team, 288, 290
Socioeconomic problems, 236
Sociopath, 249
Sociopathy, 236
Somatization disorder, 265
Somatoform disorder, 264–265, 267
Sommerness, William D., 339, 341, 343, 345
Spielberger State Anxiety Inventory, 291
Spinal disorders, chronic
  barriers to functional recovery of, 236–237
  functional restoration of, 237–241
  role of measurement in rehabilitation of, 232–233
Spinal measurements design, 234–236
Spine Education Center (Dallas), 21
Spondylolisthesis, 95
Stabilization training, 134–136
Stair climbing, gender and age as factor in, 184–188
Standing jobs, 179
Strain gauge, 73
Strength tests, pre-employment, 97–98
  functional, 98–100
  muscle, 100–101
Stress management
  education in, 13, 33
  sessions, 16
  using pause gymnastics, 37
Stretching exercises, 22
Structural (musculoskeletal) evaluation, 207, 208
Subjective functional profile, 207
Supervisors
  health education programs for, 11, 14
  importance of, in implementation of health education programs, 18
  peer pressure from, 326–327

Surgery, functional restoration program and 238–239
Surveillance systems
  guidelines for musculoskeletal injury, 41-42
  hazard, 41
  health, 40
Sustained intense work, 72
Swanbum, Norma E., 317
Sweden, 56
Swimming, 306. *See also* Water exercise programs
Symptom control development, 278–279
Symptom magnification syndrome
  background of, 257–258
  decision tree, 262, 269, 270
  developmental origins of, 258–261
  evaluation of, 269
  identification of, 262–263
  treatment of, 274–280
  types of suffers of, 264–269
Symptom magnifiers, 265
  differentiation among types of, 265
  type 1, 264–265, 280
  type 2, 265–268, 280
  type 3, 268–269, 280
Systemic fatigue, 83

## T

Tasker, D., 35
Tate-Henderson, Suzanne, 195, 311
Telecommunication networks, 49–50
Temperature, 57
Tendinitis, 76–77. *See also* Repetition strain injury
Tendon evaluation, 62
Tennis elbow, 58–59
Tenosynovitis, 43. *See also* Repetition strain injury
Tension/stress problems in neck and shoulders, 59
Test design for pre-employment screening, 102
Thoracic outlet syndrome, 59
Tool adaptation, 279

Tools
  ergonomically designed, 44
  preshaped, 61
Tramposh, Anne, 195
Transcutaneous electronic nerve simulator (TENS), 289
Trapezius strain, 304
Trauma. *See* Cumulative trauma
Traumatic neurosis, 249–250, 255
Treatment, rehabilitation at same time as, 305-307
Triage principle, 116
Troup, J. D. G., 96
Trunk mechanic work simulation, *218*
Trunk muscles
  programs to improve strength in, 46
  strength and function of, 100–101
  testing strength in, 235–236

## U

Unable clients, 349–350
Uncontrollability, 261
*Uniform Guidelines on Employee Selection Procedures*, 95
Unions. *See* Labor unions.
Unmotivated clients, 352–353
Unskilled jobs, 178

## V

Validation of pre-employment screening protocol, 103
Validity of measurement, 235
Valpar and Singer component work samples, 203
V-Code, 266
Very heavy work classification, 178
Vibrations as cause of cumulative trauma, 57
Vocational assessment, 207
Vocational expert
  role of, 175
  in Social Security disability hearings, 176–177
Vocational Preference Inventory, 203
Vocational rehabilitation, 197, 203–206

## W

Wang, M. J., 68
Warm-up phase of exercise program, 31
Water exercise programs, 288–289, 306
Weight capacities, gender and age as factors in, 184–187
Weight control, 33
Weight training, 31–32
West, Jodi Landis, 283
WEST Tool Sort, 273, 277–278
Westmoreland Coal Company, 19–20
White, A. A., III, 181
White, Arthur H., 21–22, 26, 28, 134
Wide Range Achievement Test, 203
Wilke, N. A., 198
Women. *See* Female workers
Wood, Andrew, 30, 31
Work, 54. *See also types of work*
Work capacity evaluation, 144, 145. *See also* Functional capacity evaluation
Work capacity services and return to work issues, *226–227*
Work capacity specialist, 205
Work environment improvement, 85
Work Environment Scale, 291
Work force changes, 258–359
Work function themes, 273, 277–278
Work hardening, 140
  definition of, 196, 239
  environment for, 279
  as evolutional process, 196
  for fearful clients, 350
  need for, 195–196
  overview of, 193
  review of literature on, 196–198
  therapist in, 336
Work hardening programs, 46. *See also* Comprehensive psychophysical model for work hardening
  benefits of, 191
  comprehensive psychophysical, 204–223
  generalizations regarding, 223–224
  occupational therapy in, 200–202
  physical therapy in, 198–202

problems regarding, 224
treatment of symptom magnification syndrome within, 275
vocational rehabilitation counselor, 203–204
Work pattern, 72
*Work Practices Guide for Manual Lifting*, 86, 87, 93
Work practices modification after injury, 131
Work Safe program (Australia), 56
Work simulation. *See* Job simulation
Work tolerance screening, 144, 145. *See also* Functional capacity evaluation
Worker. *See* Employee
Worker role indentification, 279–280
Worker screening. *See* Employee screening; Pre-employment screening
Worker training programs to reduce back injury, 45–48
Workers' compensation
  costs, 333–334
  employer and, 4
  due to injury, 92
  laws regarding, 314
  problems associated with, 308–309
Workers' Compensation Act, 340
Workers' compensation forms, 163
Workers' compensation insurance carriers, 328–329, 356–357
Workers' compensation system
  awards, 340
  and chronic pain patient, 183
  immediate care of work injuries and cost of, 115
  legal processes and, 339–340
  participants in 3–6
Work-injury management. *See also* Employer work-injury management,
  employer's role in, 318–322
  participants in, 3–6
  problems in, 3
  therapists' roles in, 331–337
Work-injury treatment
  acute vs. chronic, 304
  components of, 116–117
  delivery systems for, 117–119, 128–129
  directed services for, 124–125
  example of progressive, 120–124
  perks for employees receiving, 120
  physical therapy treatment for, 130–132
  trends and future challenges in, 132–133
Workplace design. *See also* Workstation design
  importance of, 15
  for sedentary work, 13
Work-related rehabilitation goals, 210–211, 305–307
Worksite
  analysis of, 86–87
  definition of, 54–55
  form for evaluating, *88–89*
Workstation design
  back injury prevention and, 85
  as cause of cumulative trauma, 157
  for control of musculoskeletal problems, 47–48
  to decrease likelihood of repetition strain injury, 84
Workstations
  in comprehensive work hardening programs, 216
  examples of, 217–222
Wrist
  conditions of, 58–59
  postures, 66, 68, *70*
Wx: Work Capacities, Inc., 206

# X

X-ray screening, pre-employment, 97, 108